Barbara Goldoftas

The Costs of Ecological Decline in the Philippines

UNIVERSITY PRESS

2006

OXFORD
UNIVERSITY PRESS

Oxford University Press, Inc., publishes works that further
Oxford University's objective of excellence
in research, scholarship, and education.

Oxford   New York
Auckland   Cape Town   Dar es Salaam   Hong Kong   Karachi
Kuala Lumpur   Madrid   Melbourne   Mexico City   Nairobi
New Delhi   Shanghai   Taipei   Toronto

With offices in
Argentina   Austria   Brazil   Chile   Czech Republic   France   Greece
Guatemala   Hungary   Italy   Japan   Poland   Portugal   Singapore
South Korea   Switzerland   Thailand   Turkey   Ukraine   Vietnam

Published by Oxford University Press, Inc.
198 Madison Avenue, New York, New York 10016

www.oup.com

Oxford is a registered trademark of Oxford University Press

Library of Congress Cataloging-in-Publication Data
Goldoftas, Barbara.
The green tiger : the costs of ecological decline in the Philippines /
Barbara Goldoftas.
p. cm.
Includes bibliographical references and index.
ISBN-13 978-0-19-513510-7; 978-0-19-513511-4 (pbk.)
ISBN 0-19-513510-5; 0-19-513511-3 (pbk.)
1. Natural resources—Philippines. 2. Sustainable development—Philippines.
3. Environmental policy—Philippines. I. Title.
HC453.5.G65 2005
333.7'09599—dc22   2004066290

Excerpt from Yehuda Amichai, "Tourists," from *The Selected Poetry of Yehuda
Amichai*, edited and translated by Chana Bloch and Stephen Mitchell. Berkeley:
University of California Press, 1986. Copyright © 1986 The Regents of the University

In the beloved memory of my parents,
Tobi and Ann Goldoftas

Preface | The Forest Guards

In January 1991, as the first Gulf War began, I flew across the world to the Philippines to work as a writer. With bombs trained on targets around the Middle East, it was an uncertain and fearful time to be traveling. In Manila, a bounty had been offered for each American head, and the day before I arrived, a powerful bomb exploded outside the U.S. Information Agency, blowing a crater into the street and bits of flesh into the trees. As I began my first assignment, traveling to remote villages to visit projects funded by a small development agency, I found that I had entered a realm of conflicts quite separate from my own country's war. The hardships humbled me, and I soon stopped thinking about my own safety. Within a few days, I was in a remote indigenous village interviewing men who felled and dragged trees through the forest with their *carabao*, water buffalo. I listened as teachers taught children to count wood in board-feet so their parents would not be underpaid. I walked through the ragged landscape of a national park in the Bicol region, where towering hardwoods had been reduced to charcoal and burned stumps.

These communities, I learned, were not unusual. Much of the magnificent tropical forests that once covered the archipelago's interior had been cleared in just a few decades, and the remaining forest was being cut. Over the course of a year, I traveled along the forest frontier, reaching far-flung villages by taking boats down rivers, waiting for daily jeepneys or the occasional motorcycle, or walking. The deforestation had happened so rapidly that people still spoke with anguish about the changes that it brought. From northern Luzon to southern Mindanao, I heard the same stories—from farmers whose rice fields had washed away in flash floods, church workers organizing ecology groups to oppose the logging, tribal leaders whose way of life was disappearing with the trees. The decline of the forests meant the decline of the logging industry, but what people spoke of, as they pointed to the rivers and hills, was the disruption of the seasonal rains, thundering flash floods, depleted soil.

The thousands of islands that make up the Philippines are divided into three main sections: the large northern landmass of Luzon and the islands that surround it; the smaller middle islands, the Visayas; and the large southern territory of Mindanao. These areas in turn are divided into more than a dozen regions, some with distinctive characters, and the regions themselves are divided into separate provinces, which over the years continue to be subdivided. There are more than 70 different languages and dialects spoken. Some of the provinces and islands differ markedly in their language, cuisine, and political character, and for Filipinos and visitors alike, working in the archipelago means continually crossing these boundaries. Until the late 1980s, when the new democratic era created a political opening, questioning the logging and pushing for conservation were dangerous, futile work.

In 1991, the young environmental movement was fortified by two events. In early November, a typhoon crossing the Visayas hit the coastal city of Ormoc, Leyte, and mudslides catapulting down the logged-over foothills killed thousands of people in minutes. Villagers in remote areas knew that floods and mudslides visited clear-cut hillsides, but what happened at Ormoc showed the rest of the country the harm that ecological havoc could bring and the way that common people bore its brunt. A few weeks earlier, Nerilito (Neri) Satur, a young priest known for defying illegal loggers, had been brutally murdered in Mindanao. Father Neri's murder sparked a resolve, and other religious vowed to continue his work. Although journalists covering logging had been receiving death threats, for weeks the national newspapers featured stories about Father Neri and the role that logging played in Ormoc's tragedy.

I used to say that I first went to the Philippines for adventure but also out of obligation. I felt almost a duty to write about a country whose economic and military histories were so closely tied to those of my own country. Like other young western journalists working overseas, I gravitated toward places bound by fewer rules, which could bring exceptional access at every level of society. I now understand that my impulses were also deeply personal. I was born to half a family; the European side was almost obliterated in the slaughter of World War II. A loss like that takes a lifetime to grasp, but I have recognized that, raised with stories of the missing relatives and my father's flight through Europe in 1940, I was drawn to learn about conflicts in other societies and the ways that they mark a person, a family, a country.

I went to the Philippines intending to stay for three months and stayed for a year. From my base in Manila, I wrote about stories ranging from the economic implications of Mount Pinatubo's catastrophic eruption to the country's dependence on income from Filipinos working overseas to the thousands of prostitutes around the U.S. military bases. Although trained as a plant ecologist, I had not intended to write about the environment, but I found myself increasingly disturbed by the destruction in the countryside and the disorder it brought to rural communities. As I reported on migration, overseas workers, the widespread prostitution, the spread of HIV, the militarization of the countryside, I realized that they were different parts of the same puzzle and could not be separated from the growing ecological crisis.

In the Philippines, the story of logging is also the story of fishing and mining, all of which are rooted in an ownership of land that is highly concentrated, leaving millions of farmers landless. The rural populations, although dependent on the once-plentiful natural resources, were recklessly exploiting them. This overuse deepened the landlessness of the poor, who also bore the brunt of the countryside's decline. The result was millions of ecological refugees who flocked to coastal villages, where they added to the pressure on the fisheries, and into the forested uplands, where they cleared land to farm. They also crowded into the cities, usually as squatters living illegally on open land. In this way, the rural disarray deepened the problems of the urban environment, which in Manila included congestion that slowed economic activity and daily life, inadequate supplies of water and power, and air pollution that darkened the air, at rush hour obscuring the view across a busy street.

After I left the Philippines, I could not forget the bravery and fear I had witnessed along the forest frontier; the spare fishing villages; the ecological refugees; the migrants cutting the forests, plot by plot. I also thought about Father Neri. A few weeks after he died, I visited his monastery in Valencia, Bukidnon, and stood in his small room. Although I never met him, he lodged in my mind, raising question after question that I could not answer. Those questions stayed with me, and I came to see that they could be asked about countries around the world where there were conflicts over trees and oceans and mineral wealth. Four years later, I returned to try to find some answers. I spent the rest of the 1990s returning, traveling from island to island.

It was a time of profound political and economic change in the country. There was less of the unruliness, which I had come to recognize as a sign of weak governance and which no longer drew me. Instead, I wanted to understand what role it played in the ecological decline. I wanted to learn why the ecological damage had been so dramatic; what fueled it, domestically and internationally; what its political, economic, and social consequences had been; and what an environmental movement born of priests, peasants, businessmen, and the elite had accomplished. I wanted to know if there had been warnings about what the costs of ecological decline might be and, if so, why they had been ignored. I wanted to know what was being done, in what direction the country was headed.

For decades, the Philippines' economy foundered as its neighbors grew into "tigers" whose success was noted worldwide. In the early 1990s, President Fidel V. Ramos promised that the country, long considered the poor relation in the region, would join them. The streak of Asian tigers had developed with little regard to the environmental destruction that the economic growth brought. Rather than adding the burden of industrial waste to that of the overused resources, vowed Ramos, the Philippines would set a model as a "green tiger" and develop in a manner that was more ecologically sound. The economic uncertainty of the late 1990s and early 2000s has slowed the country's efforts to bound ahead on tiger claws. Still, the prospect of becoming a green tiger and dealing with degraded resources and addressing environmental concerns during development, rather than afterward, remains a goal.

This book is organized around the country's main ecosystems, an order that reflects a plant ecologist's view of the world as well as the Philippine environmental movement's work. Both the government and nongovernmental organizations targeted first the "green" sector, or forests, the first ecosystem to be overexploited; then the "blue" sector, or coasts and fisheries, which were next overused; and finally the "brown" sector, the urban and industrial environments. The chapters on the forests give some of the history of the use of natural resources in the Philippines. The section on the coasts visits models created to manage the reef fisheries and the alternatives of ecotourism and industry. The section on land and the urban environment looks at how the patterns of land use laid a foundation for the use of other natural resources and how in urban areas, problems of human health are the next environmental frontier.

Much of the 1990s were a time of hope in the Philippines. That hope is reflected in these pages, which tell stories of communities confronting the consequences of degraded natural resources. Some villages and towns were struggling to conserve and manage what was left; others, to find economic options that would not destroy their communities; still others, simply to endure. To witness their efforts was instructive and sobering, and it was a privilege. I tell their stories partly because they show clearly the extraordinarily complex nature of serious problems being faced in countrysides and growing cities around the world. Some of their stories also capture pivotal moments in the country's environmental movement. The lessons learned and the strategies developed there, anchored in communities and bolstered by national policies and growing political will, also offer useful models that have succeeded else-where. More than a decade after I first arrived in Manila, these efforts are more important than ever.

While traveling, I relied on local people to escort me and translate for me. In militarized areas, they kept me safe; in safe places, they interpreted for me, fed me, opened their homes to me. Sometimes I identify them; sometimes I protect their identity. I am grateful to them all. My hosts often asked about my life and perspective; just as they opened a window into their world, I offered a window to another life they otherwise might not have encountered. In most places, I was usually asked the same question: "Do you eat rice?" From their experi-ence with other foreigners, they assumed that I preferred American food to Filipino food—salty and sour, with piles of rice. The question asked about more than food, though, and from my answer, they could tell not just that I eat—and enjoy—Filipino food, but that I did not find the country itself to be lacking.[1]

Occasionally, Filipinos would tell me that I was very "American"—which usually meant direct, unwilling to resign myself to uncomfortable situations, insistent that problems could be solved. In a culture that prizes indirectness, some people welcomed these qualities while others found them jarring. Some also implied that I harbor optimism where little might be warranted. I partly credit my Lithuanian father for the optimism and can-do attitude, but I also have been influenced more than I will ever know by the years I spent trying to get to know the Philippines. I learned about fortitude, determination, and prob-lem solving from the many people of all backgrounds whom I met, observed, interviewed, listened to, accompanied, grew fond of, stayed with. The Philip-pines is sometimes dismissed as a worst-case scenario with a future burdened by both corruption and ecological degradation. It is also, despite its continued political turmoil, a place where new generations, refusing to bow to that label, continue to try to solve their country's greatest challenges.

Acknowledgments

No book could be written about a country across the world without the contributions and generosity of countless people. I would like to extend my thanks to everyone who, over the years that I worked on this book, helped make it possible; inadvertent omissions do not indicate a lack of gratitude. All errors within these pages are entirely my own.

Nan Levinson was as steadfast, astute, and tireless a writing companion as any writer could hope for. Caroline Chauncey, Marcia Goldoft, David Levine, and Sandra Steingraber generously read the entire manuscript, offering wise comments and encouragement. For reading individual chapters or parts of chapters and lending their insights, I would like to thank Jennifer Ackerman, Richard Anson, Danny Balete, Jun Borras, Will Combes, Kirsten Engel, Pinky Esguerra, Emilyn Espiritu, Jenny Franco, Lisa Goldoftas, Bebet Gozun, Larry Heaney, Tony La Viña, David Luberoff, Ipat Luna, Coke Montaño, Marianne Moore, Sally Ness, Nick Rodenhouse, Charles Rubin, Laura Van Dam, and Alan White.

In the Philippines, I learned the meaning of hospitality; each chapter rests on the contributions of more people than I could name. In every town, province, government agency, organization, and urban neighborhood I visited, I owe huge thanks. I would particularly like to acknowledge Perry Aliño, Chip Barber, Sheila Coronel and the Philippine Center for Investigative Journalism, Beng and Butch Dalisay, Mercy Fabros, Luis Francia, Myrna Francia, Eugene Gonzales, Bebet Gozun, Tony La Viña, Viking Logarta, Ipat Luna, Helen Mendoza, Coke Montaño, Tony Oposa, Mô Ordoñez, the Philippine Center for Policy Studies at the University of the Philippines, Laurence Pochard, Vaughan Pratt, Mary Racelis, Boyette and Luz Rimban, Joel Rocamoro and the Institute for Popular Democracy, Howie Severino, Suzanne Siskel, Dinky and Hector Soliman, Raquel (in fond memory) and Rigoberto Tiglao, and Marites Dañguilan Vitug. For their sustained friendship over the years, to Orville and Pia Solon and Manny and Pinky Esguerra I surely owe an *utang na loób*.

I have been lucky to blend a life of writing with one of teaching. For their artistic and intellectual companionship, I would like to thank colleagues at the Massachusetts Institute of Technology, particularly those in the Program for Writing and Humanistic Studies and Philip Khoury; and at Wellesley College. Both institutions generously supported the research for this book. As a visiting writer, I spent a productive semester at the School for Critical Studies at the California Institute of the Arts. I was privileged to spend two wonderful years at the Mary Ingraham Bunting Institute, then at Radcliffe College, and I would like to acknowledge Rita Nakashima Brock, Renny Harrigan, and my fellow fellows. Through the Bunting, I worked with the Radcliffe Research Partnership Program, and I am grateful to Leslie Bennett, Carrie Jenks, Jessica Leino, Cynthia Lin, and Vicky Obst for their enthusiasm and careful research. At the Boston University School of Public Health, I thank faculty and students alike for their patience and wisdom.

For their early encouragement and prodding to write this book, I am indebted to Jennifer Ackerman, Bill Allen, Alan Lightman, Michael Mandel, and Seth Shulman. I would also like to acknowledge George Packer and Paul and Edna Hutchcroft. For our conversations I would like to thank Kirsten Engel, Tim Forsyth, Hugh Gusterson, Sue Helper, David Luberoff, Jethro Pettit, Scott Saleska, and Jonathan Schlefer. For helping make it possible for me to complete the book, I would like to thank Leonid Gordin, Rebecca Herrmann, and Debra Shapiro. I wholly admire the vision and photography of Marissa Roth and Ryan Anson.

I am grateful to agents Charlotte Sheedy, for her interest in this book, and Neeti Madan, for seeing it through. At Oxford University Press, I would like to thank Joyce Berry for seeing promise in the book proposal and Susan Ferber for her unflagging support, guidance, and skill as an editor. I could not imagine a more dedicated and reassuring production editor than Stacey Hamilton.

For their general support and belief in me, thanks are not enough for Marcia Goldoft; Lisa Goldoftas and Hooman Aprin and my beloved niece and nephew, Dora and Joey; Neil and Gail Goldoftas; Joyce Hackel and Robin, Gabriel, and David Lubbock; Rae Weil; and, of course, Ann and Tobi Goldoftas.

Contents

Glossary

amihan northeast monsoon

ayaw ko, ayaw namin I don't want it, we don't want it

baboy pig

bago new

bahala na come what may; a feeling of fatalism

bakit why

banca small boat

bantay dagat watchdogs of the sea

barangay village

barrio village

batô stone

bayanihan a spirit of cooperative work; camaraderie

bolinao anchovies; herring

bolo machete

buhawi waterspout or tornado

CAFGU Civilian Armed Forces Geographical Unit (an armed paramilitary group)

camote sweet potato

carabao water buffalo

coco coconut

cogon saw grass

datu tribal leader

DENR Department of Environment and Natural Resources

dipterocarp a family of tropical hardwoods with a double-winged seed

Dumagat an indigenous people in Luzon

estero canal

Falcata a fast-growing species of tree exotic to the Philippines

FARMC Fisheries and Aquatic Resources Management Council

galang respect for hierarchy and authority

Gmelina a fast-growing species of tree exotic to the Philippines

gulay vegetables

habagat southwest monsoon

hanapbuhay livelihood

hectare 2.47 acres

hiyâ shame

Hukbalahap (Hukbo ng Bayan Laban sa Hapon) People's Army to Fight the Japanese; Huks

indios Spaniards' term for Filipinos

jeepney public utility vehicle, a cross between a truck and van

kaingin shifting agriculture

kaingiñeros farmers who practice shifting agriculture; often used to describe all farmers who clear forested land to farm

kalaw a hornbill (type of bird)

kamagong Philippine ebony

kinaiyahan environment

lahar liquefied volcanic ash

lapu-lapu grouper (type of fish)

lechon roasted pig

lumad indigenous peoples in Mindanao

Malacañang the presidential palace

milo civet cat

Moro Muslim

muro ami drive-in net fishing

naninibago to get a sense of a new place

nilad a kind of water lily

nipa a kind of palm found in brackish swamps

novicio priest in training

pakikisama the ability to get along with others; camaraderie

pancit noodles

PCSD Palawan Council for Sustainable Development

poblasyon town center

sangguniang bayan municipal council

sangguniang panlungsód city council

sari-sari general store

sitio community

tanglad lemongrass

tanguinge Spanish mackerel

trapo traditional politician; old rag

tsismis gossip

utang na loób debt of gratitude; debt of the heart

wala none; without

South China Sea

Cordillera

Sierra Madre

LUZON

Bolinao • • Baguio

Lingayen • • Dagupan

Philippine Sea

Mt. Pinatubo ▲

SUBIC BAY • Quezon City

☉ *Manila*

Dasmariñas • *LAGUNA DE BAY*

B I C O L

THE PHILIPPINES

MILES

0   50   100   150

MARINDUQUE

MINDORO

MASBATE   **SAMAR**

VISAYAS

PANAY   Ormoc •

El Nido •   Iloilo •   • Sagay   **LEYTE**

 • Taytay   • Bacolod **CEBU**

                        • Cebu

NEGROS   **BOHOL**

PALAWAN • Puerto Princesa

Dumaguete •

Apo Island •

Sulu Sea

 • Cagayan de Oro

Kitanglad ▲▲ • Malaybalay

Range •   • Valencia

MINDANAO

Davao •

Zamboanga •   Digos •

Malalag • *DAVAO GULF*

SULU ARCHIPELAGO

ROBERT LEVERS

1 | The Pearl of the Orient Seas

The first time I went, Mindanao seemed a world apart. It is the second largest island in the Philippines and lies about 400 miles south of the capital. Mountainous over much of the interior, it has a meandering coastline that protrudes in all directions, eventually breaking up into its own small archipelago that scatters west across the Sulu Sea. In the early 1990s, traveling from Manila to even nearby rural provinces felt like backing up in time, and this was particularly true in Mindanao. Outside of any sizable town, the fittings of modern life quickly dropped away: first the larger buildings and vehicles, then the telephones, and finally the water pumps and wires carrying electricity, leaving only the road until, a rutted maze of stones and dirt, it, too, little resembled a modern convenience. Villages consisted of at most a cluster of houses lining dirt roads or footpaths and a one-room store. Even the *sari-sari* store, named after a Tagalog word for variety, offered few manufactured goods other than the basic necessities, and they were usually the same basic necessities—rough soap and toilet paper, sweets, packets of shampoo and conditioner, plastic water dippers, skinny tins of fish and tomato paste, matches, and glass bottles, scuffed from reuse, of Coca-Cola and San Miguel beer. Cosmopolitan Manila, with its crowded streets, high-rises, and government offices, seemed impossibly far away, and the trip might require up to a day's travel, for those who had the means. Outsiders, including government officials, rarely came, and foreigners were so uncommon that in my white face and long nose, children would see a ghost or devil and wail with fright.

It was all too easy to encounter a place like Mindanao and think it unchanged. Portions of its landscape resembled grainy, gray photographs taken by western explorers more than a century ago. Fishermen lived in small wooden houses erected on stilts over the water. Farmers prodded *carabao* through the mud of flooded rice fields. Beside the dirt roads that ran toward the mountains, families raised animals under their houses. The only routine links between the more remote *barangays*—villages—and the closest town market were the jeepneys, the colorful cross between a pickup truck and commercial van used to haul passengers, produce, and livestock alike. Officially called public utility vehicles, jeepneys were first created after World War II from the motors of Jeeps left behind by the U.S. Army. With their windows hung with bright tassels, their dashboards adorned with religious statues, and their sides painted with names and slogans, the jeepneys brightened both congested city streets and rural routes, where they were often the only large vehicles on the roads.

The Philippines is a place of enormous natural beauty. More than 7,000 islands make up the archipelago, most of them unnamed and uninhabited. Originally the islands were ringed by mangroves, coral reefs, and rich fishing grounds, their mountainous interiors thickly forested and crossed by wide rivers. They bore plants and animals that were unusually abundant and varied. New species form in isolation, such as on mountains and islands, and although the total land area of the Philippines is about the size of Italy, the islands once supported a flora and fauna with nearly unparalleled diversity.

. . .

Despite its seeming distance from modern life, Mindanao, like much of the archipelago, had been transformed as the natural resources that once sustained the rural populations were quickly depleted. The forests were logged, the coasts overfished, the coral reefs eroded. Large-scale commercial logging, which began on Luzon and in the Visayas, had reached Mindanao by the late 1960s. Within three decades, little forest remained, and its retreating edge was a chaotic place.

The forest frontier was unstable both ecologically and socially. Remote areas tended to be fearful places fractured by violence. Bans on cutting in primary forests and exporting logs together with a huge reduction in logging licenses had ended most legal logging, but with little enforcement of environmental laws, the cutting continued illegally. As the most lucrative economic activity around, it drew those in power, including the military, and challenging the cutting could be daunting and dangerous.

The Philippines had drawn international attention in 1986 for its relatively peaceful change of power from the dictatorship of Ferdinand Marcos to the elected government of Corazon Aquino. Outside its borders, the country was seen as a promising new democracy, but the view from inside was different. The "people power" revolution that ousted Marcos had shifted power primarily in Manila. Places such as Mindanao remained semifeudal. Local politicians ran villages, towns, and even small cities like warlords, dominating political and economic activity and maintaining control with private armies. Civil war split many rural provinces. The New People's Army—the armed wing of the Communist Party of the Philippines—remained strong, as did the military's anti-insurgency strategy of "total war." In a predominantly Catholic country, about 5 percent of the population were *Moro*, Muslim, and lived mostly in extremely poor provinces in Mindanao, where two guerrilla groups fought for independence. Kidnapping seemed endemic, and the conflicts, generally referred to as the "peace-and-order situation," continued to dislodge whole communities. Even in places where combat had receded, many refugees had not yet gone home, but life had begun to return to normal, the kind of halting normalcy that follows war.

The transition from dictatorship was long and fraught with political and economic uncertainty. During its six years, the Aquino administration defended six coup attempts by supporters of Marcos and a rebellious wing of the military. The country owed more than 27 billion dollars to foreign lenders, and the government relied on new loans and grants to keep afloat.[1] The economy, having slumped into negative growth at the end of martial law, was in shambles. As the government took stock, it also became clear that the widespread destruction of the archipelago's extraordinary natural environment—a once-vast endowment that had both enriched the elite and sustained the rural poor—constituted an additional economic burden.

With few formal institutions at the local level, clergy typically offered the only leadership as communities tried to stem the downing of their forests. This was true in Malabog, an upland town about 45 miles from the center of Davao City in southern Mindanao. An ecology committee organized through the church had begun to confront the logging in the hills, which I went to investigate late in 1991 on my first trip to Mindanao. From the coast of the immense Davao Gulf, the city spreads inland deep into the foothills. Like other provincial cities, Davao covers a large area, but its center was barely developed then, its market just baskets of produce arrayed on the ground. On an island known for tumult, Davao was a rough place whose mayor bragged about having rid the city of criminals by pushing them from a helicopter midflight. The militarization extended into the countryside, and in Malabog, curfew began at 10 o'clock.

Reaching Malabog from the center of Davao required most of a day and several forms of transport. After a long jeepney ride, I was lucky to get a seat on a motorbike headed there. One man perched on the handlebars while another passenger and I squeezed behind the driver, clutching our bags during the jarring, hour-long trip. As the driver negotiated ruts deep enough to swallow his front wheel, we passed by a banana plantation, unpainted wooden houses, children pumping water. From the back of the motorcycle, I could see the mountains of blue, green, and gray rolling to the horizon. Some of the foothills lay beneath the mountains like interlaced fingers, the folds in the earth visible where they had been cleared of trees. Close to Malabog, many of the mountains also were bare, and on the peaks overlooking the town, open patches in the forest revealed that loggers continued to cut.

As was usually true in remote areas, I did not know what to expect. There had been no way to communicate with anyone beforehand, and so much was unpredictable: the travel, the peace-and-order situation, the people I could interview. The town was spread out, with a short strip of concrete road, a covered common pump, and a church and convent. As the driver wheeled in the motorbike, a group waited, children in arms and underfoot. When I introduced myself, a young woman with penetrating eyes and one arm missing above the elbow asked where my "companion" was, a common question because Filipino women rarely traveled alone. Life was casual in Malabog. A woman pointed out the priest working with the ecology committee, who stood in his shorts some distance behind the church, pouring water over himself as he bathed. His back glistened in the late afternoon sun.

During the trip, the motorbike had passed people clustered on porches. It was All Souls' Night, when people gathered to commemorate the dead. Father Max—I did not learn last names in Malabog—invited me to join him in celebrating the festival. I found out then that the priest who had set up the ecology committee had been assigned elsewhere and Father Max had arrived only the month before. My actual host would be 38-year-old Dodong, a lay churchworker and one of the committee's leaders. Stocky and outgoing, Dodong knew only rudimentary English, but he more than compensated for his vocabulary with his curiosity and sincerity.

We had dinner in the house next to Dodong's at a table that took up much of the room. We ate the standard fare for festivities: ground corn, potatoes and peas, peas and goat, *pancit*—rice noodles with vegetables—and *lechon*—roasted pig. Dessert was sweet wedges of sticky rice with banana and cassava pudding sprinkled with ground peanuts. As we ate, the sky turned black. Even in the lit house, I was a spectacle. Boys crowded against the walls to stare, and other faces pressed up to the windows or peered around corners. When I laughed a loud American laugh, mimicking echoes followed.

After the table was cleared, young men brought guitars and sang mostly British and American oldies, laboring to strip away their Filipino accents. Guests sang, too. When my turn came, I sang a biblical song in Hebrew and English. "And every man 'neath his vine and fig tree shall live in peace and unafraid." I chose the song thinking that religion might offer a way for my hosts to get to know me, and it did. Dodong asked what it meant to be Jewish and how I celebrated *fiestas*—whether people sang and danced, whether we served a lot of food. Someone then inquired why the U.S. public had elected particular presidents, what I thought of the current president, how his policies might affect the Philippines. We did not speak directly of the militarization and violence in Malabog, nor would Dodong explain why there was a curfew, although he would reveal later that the churchworker I met earlier had lost her arm and her husband while in the New People's Army. We did, however, speak of politics and logging, the other common denominators in Mindanao and everywhere along the forest frontier.

Father Max believed that the previous priest had been assigned to another parish because of his environmental activism. The ecology committee had spurred people to question the logging. "The situation became too intense," he said. "The bishop told me to provide support but not be the leader. That was the first advice I got." As he began working with 100 lay leaders in more than 30 chapels, they worried that if he got overly involved with environmental issues, he also would be "pulled out," he said. "They're afraid that they won't have another assigned here." He admitted that he, too, was afraid, but he felt a responsibility to help the community oppose the logging: "People say the economic base has been affected, that the climate has changed, the land isn't as productive."

Dodong did not hesitate to voice his frustration. Like others in upland communities that most felt the effects of deforestation, he spoke with anger of "defending the forest." The Philippines, he said, "is a rich country, but it's being exploited to the hilt. The logging companies will consume what they can and then leave." He knew that conserving the forests required the government's assistance but that both national policies and corruption had fueled the logging, at times even accelerating its pace. Most of the country's forestlands were publicly owned and managed by the national Department of Environment and Natural Resources (DENR), which was blamed for their loss. Widely faulted for extreme corruption in both Manila and the rural provinces, the DENR had also failed to replant the deforested areas, despite ample international funding.

"The DENR was created to protect the environment," said Dodong, "but they also let the big companies cut the trees. They tell us that it's only selective cutting, but that's not what the companies do. They usually cut without regard to small trees, and they use equipment like the Skagit—the yarder—that smashes everything in its path. I'm committed to defending the forest; if the last stretch is cut, the rivers here will dry. I want to tell our fellow brothers in the

U.S. that in at least one part of the Philippines, people are fighting for the environment. If we can't win this time, maybe the next generation will."

There were two places I could stay in Malabog—the convent or Dodong's home with his young wife and five children. Dodong mentioned that the last foreigner to visit, an American filmmaker, had so worried about her safety that she slept in the convent and refused to drink the water. Filipinos value hospitality; refusing an invitation can insult. I accepted. His house was made of wood on stilts, like those the motorbike had passed. It had one bed draped with mosquito netting in a separate room. The kitchen consisted of a table, gas burner, and sink. After settling the children on the floor in the main room, he secured me under the net, and, telling me not to worry, left for the night watch. The night grew noisy with crickets and panting dogs and the unfamiliar sounds of what must have been goats or pigs beneath the house. I did not hear the gunfire that often punctuated the night in militarized zones. Eventually the pounding of rain began. I wrote until the thick white candle lighting my notebook burned down an inch and then blew it out. In the morning, we drank sweetened coffee in the simple kitchen and crouched beside the window.

To understand the disruption the logging brought, I had to go to the uplands, so we walked high into the hills to an indigenous village to meet with several *datus*, tribal leaders, who worked with Dodong's ecology committee. During the day-long visit, Dodong translated. We talked for several hours in a bamboo hut, then had lunch and a nap. Before we lay down, one of the *datus* played a native guitar—a long and thin instrument—and sang a rhythmic, nasal song. Although the men laughed about his serenading me, I sensed that they were trying to include me so that I might leave with a respect for who they were and how they lived.

A 38-year-old *datu*, whose father also had been a chieftain, talked at length about the logging and how it had changed life in the indigenous communities. "Before, this place was full of trees, and the forest was here. We were innocent to what logging would do to us. After the companies came, settlers came and cut trees. The tribals were displaced. Because of our ignorance of the equipment and the people who came, we were afraid. It's our custom not to be near Christians—Visayans—because we can't understand what they are saying. So we fled into the mountains." Often described as shy, indigenous people would retreat from strangers, whom he called Visayans after their main language. Further north, lowlanders would similarly be referred to as Tagalogs.

The village had shrunk from 100 families to 26, their territory from about 1,200 acres to 30. The logging had depleted the forests that once supported the community but had brought no alternative. "Before, there were wild animals for food," said the *datu*. "We would hunt deer, monkeys, lizards, wild chicken,

*Roxas Boulevard
along Manila
Bay.*
MARISSA ROTH

and *baboy, kalaw, milo* [wild pig, bird, civet cats]. There were also many fishes and big frogs. There was fruit in the forest. We could get a kind of salt from the trees. There was an unending supply of water. Without the forest, we have suffered. There's no fresh fish, there's no meat, and our farms produce much less. It's also harder to get water. Most important, the soil isn't fertile anymore. We used to make bags from rattan to sell at the market. Now we can't do that either. We work as farm laborers for pay, but we don't earn enough, only about 25 pesos [a dollar] a day. We think we will face hunger. Some people retreated into the forest that remains. We stayed because this is our land and we're tired of moving to other places. We're hell-bent on protecting this piece of land." Safeguarding the land meant challenging local corruption, which was risky. Before I left, one of the *datus* said that members of neighboring tribes—whose leaders allegedly were accepting bribes from the loggers—had "threatened that there will be a headhunting season if we continue speaking out on the environment."

. . .

In the late 1800s, in a poem penned before his execution in Manila by a firing squad of Spaniards, the martyr José Rizal sentimentally called his country a "region of the sun caressed" and "pearl of the Orient seas." In addition to the natural beauty of the coasts and forests, the islands also had an agrarian beauty. Even, flooded fields of rice covered the few valleys and flat coastal plains. In the northern provinces, without enough flat land to farm, farmers centuries ago carved the hillsides into narrow shelves stepping up the slopes. These rice terraces reshaped those parts of the northern provinces and are revered for their beauty and for the hard work, ingenuity, and devotion to a place that they represent.

Filipinos still call both the country and Manila the "pearl of the Orient," and as recently as World War II, the countryside resembled the tropical paradise that its nationalist writers so tenderly evoked. In slightly more than one generation, before their species could be counted and identified, most of the country's primary forests were leveled, leaving open valleys and uplands that are commonly and accurately described as bald. The deforestation rate then was one of the fastest in the world. Despite laws requiring commercial loggers to replant, they rarely did, and the roads they built opened the forest to illegal small-scale cutters and migrant farmers, who cleared the forest that remained.

By the early 1990s, the ecological consequences of the logging were all too visible. In some hilly areas logged years earlier, soil had eroded or been stripped of nutrients by a few years of farming, and the land lay barren. In newly cleared foothills, without tree roots to anchor it, the deep, loose top layer of soil fell away in armfuls. In the valleys, rivers ran low, their beds studded with dry boulders. Like the forests, the country's reefs and offshore fisheries had been heavily degraded by a combination of large commercial operators and subsistence fishermen. As the industry expanded, fishermen worked longer hours and traveled to farther waters to maintain their catches. They also relied on destructive fishing methods such as using dynamite and poison, which boost catches while indiscriminately killing corals, fish, and other ocean life.

Although the total number of species once found in the Philippines will never be known, biologists estimate that the combined landmass of nearly 116,000 square miles supported about 13,500 species of plants and more than 1,000 species of animals. These numbers are hard for those of us from the species-poor North to comprehend. Even more unusual—and important biologically—a huge portion of these species were endemic, or unique. As many as 40 percent of the country's flowering plants, 44 percent of the birds, 64 percent of the mammals, and 70 percent of the amphibians lived only on this archipelago, often on just a few islands.[2] High numbers of endemic species are one reason that island ecosystems such as the Philippines are considered ecologically fragile; when disrupted by fire, overfishing, or clear-cutting, they cannot necessarily regenerate quickly, and some will never return to anything resembling their former state.

Planting rice. Pampanga, Luzon.
MARISSA ROTH.

The Philippines still harbors some of the world's most biologically complex ecosystems, but they remain only as remnants whose full restoration is unlikely. The pressures to log and fish continue, fed by intractable corruption, greed, and poverty. Much of the ecological damage is irreversible. Across the uplands, the clear-cutting has brought the disorder repeatedly described to me in Mindanao: flash floods, deeply eroded soil, and shifts in rainfall, including droughts. The destruction of coral reefs, which shelter young fish, has depleted the populations in nearby fisheries. The loss of the mangroves that once fringed islands has similarly removed a key ecosystem that supported coastal fisheries.

In deep waters, overfishing has altered the species that the fishermen catch, suggesting that marine ecosystems are breaking down.

The reckless exploitation of the forests, reefs, and fisheries has eroded ecosystems, erased whole species, and brought enormous ecological conse- quences—not only for the fragments of indigenous communities still living in the upland forests, but for the entire country. The demise of the forests meant the demise of the Philippine logging industry, which, although immensely profitable in its heyday, along with the ecosystems that it obliterated eventually collapsed. Communities lining the archipelago's lengthy coastline have been impoverished by the decline of the fisheries, which helped support unskilled, landless laborers and their families and also provided most of the protein in their diets. The degradation of the forests and fisheries deepened the poverty in the provinces, which in the 1970s and 1980s helped accelerate the country's overall economic decline. Fleeing both ecological degradation and the persis- tent military conflicts, growing numbers of refugees moved to the cities, where they built squatters' settlements that lined waterways and covered empty lots. They also fled to less-populated rural areas. Along the coasts, migrants became fishermen. In the forested hills and mountains, where they pushed indigenous people even farther into the uplands, they cleared land to farm, in their des- peration tilling slopes so steep that it was hard to stand. By the early 1990s, 15 to 18 million people out of a population of about 60 million were believed to be living in the uplands and cutting the remaining forests, and central Mindanao was the fastest-growing part of the country.[3]

✦

The 1990s were a pivotal time in the Philippines. In the rural provinces, the polit- ical map was changing. Rural villages were still remote, and it took an enormous effort for government officials and environmental workers to reach areas where they were needed. As the democracy begun in Manila filtered to the provinces, however, warlords and longtime politicians were losing local elections to new mayors and other local officials more likely to understand and more willing to acknowledge the economic consequences of ecological change. With little expe- rience or skills and few precedents in the management of natural resources, the new generation of politicians looked to curtail the illegal use of forests and fish- eries and regulate access to them—at least until other paths toward development might be found. Under the presidency of Fidel Ramos, officials and government workers at the DENR and other agencies were talking about sustainable devel- opment, green tigers, good governance, and ways that communities could par- ticipate in decision-making about industrialization and conservation.

Environmental problems reveal societies' fault lines: the ways that they fail to soften the clout of the most powerful, or fail to protect their weaker members, or put short-term economic goals before massive environmental destruction. In the Philippines, the economy has remained skewed toward a small elite. Historically, economic activity that favored this elite—from plantation agriculture to logging—did not broadly benefit the larger population, leaving millions of people impoverished, especially in the rural provinces. Throughout the second half of the 20th century, foresters warned that to protect both the watersheds and the timber industry, the forests should be conserved—and during the same period, ample evidence showed fisheries to be declining. The government and private sector acted to safeguard neither. Instead, the ecological decline deepened the poverty in the provinces and created hundreds of thousands of refugees. These refugees in turn contributed to the ecological degradation of other rural areas and, eventually, urban areas. Overall, the widespread destruction of natural resources harmed rural economies and ultimately affected the national one.

The ways that societies approach their environmental problems also can reveal their strengths. The environmentalism that developed in the West is not monolithic. It has emphasized the conservation of natural ecosystems to protect them as wild lands or manage them for long-term use. It has worked to safeguard human health. It has bolstered the government regulation of industrial waste. It has often been controversial, and its gains, including much-needed water and sanitation systems, typically faced considerable resistance, as efforts to curb emissions and shelter wilderness areas continue to face today. Despite the considerable accomplishments of the western environmental movements, including cities able to sustain massive economic growth, critics depict it as opposed to development. Skeptics also fault it for not valuing people enough and ignoring economic need, for romanticizing traditional ways of life, for being pessimistic and focusing on problems rather than solutions, for predicting ecological crises that never materialize.

In the Philippines, the ecological crises are not imagined—mudslides, erosion, unreliable water supplies, declining fish catches, staggering levels of urban pollution. Those who oppose logging the country's remaining trees or irresponsible land use or commercial fishing that could undermine the country's future food supply are hardly exaggerating problems—they are hoping to solve them.

Environmentalism in the Philippines could be said to have grown from need and optimism. Many environmentalists originally came from the business or religious sectors or from the movement that opposed martial law. Some were influenced by the ideas and gains of western environmentalism, others moved by the deprivation in their country. They understood that the rural communities depended on a deteriorating base of natural resources and that the large-scale exploitation of these resources usually benefited the elite, not the local populations. They brought to environmental work a conviction that development that accommodated both economic and ecological considerations would better serve their country. Efforts to address the needs of refugees, urban poor, and indigenous peoples in the uplands constituted less of a romantic view of these marginalized populations than a recognition that without alternatives, they would only heighten the pressure on the natural resources. Said an NGO head, as he showed me the steep, tilled hillsides above his coastal town in southern Mindanao, "If you stop one farmer from planting, another will take his place."

The first time I went to Mindanao, with its remoteness and lack of development, it seemed a world apart. In fact, the situation there turned out to approximate that of other islands in the Philippines and other countries in comparable stages of development. In the dozen-plus years that have passed since I first witnessed the newly barren hills in Mindanao, the tensions that once seemed an artifact of that particular place and time have surfaced around the world. Bombs and kidnappings are no longer confined to places split by protracted war. The needs and demands of Muslims are no longer deemed peripheral. Similarly, conflicts over natural resources have become part of a broader public debate.

As the world's economies draw closer, it is tempting, amid the rhetoric of globalization, to think of environmental problems and the movements to address them as solely global. While some environmental problems clearly need to be addressed on a global scale, others are distinctly local and require both local and national solutions. The environmentalism that developed in the West is distinctive, and it does not necessarily transfer well to other cultures and other nations. This is the case for many industrializing countries, where ecosystems can be less resilient and where natural resources still support rural populations, remaining battlegrounds between small and large users, between the poor and the rich. In many of these countries, governments are weak and cities mushroom to extraordinary sizes before public health and environmental measures can be put in place.

As in the Philippines, many countries need an environmentalism that treats environmental protection as an ecological *and* economic problem and that could guide communities in managing the resources upon which they depend. They often need an environmentalism that suits their countries' specific ecological conditions and does not require scientific and technological prowess. They need an environmentalism that confronts the full implications of land ownership and use. They need an environmentalism that can accommodate entrenched poverty and help strengthen a government unable to fully enforce its own laws. To live in the Philippines is to see the costs of loggers, fishermen, miners, and farmers who were unconstrained by conservation. It also is to see a country rebuilding itself after a long dictatorship and war, trying to remake its economy and repair vast ecological damage, pressing for better governance, searching for solutions.

Part 1

The Forests

2 | The Lost Eden: *Noli me Tangere*

Crossed by a typhoon belt and underlain by two major tectonic plates, the Philippines lies within a geological area aptly called the Ring of Fire. Some of the archipelago's islands form the tips of volcanoes stretching up from the ocean floor, while others edge huge ridges, the spines of a range arcing north from Indonesia to Japan. These massive submerged mountains were created as the Pacific and Eurasian plates ground together millions of years ago. They still grind together, triggering earthquakes and volcanic activity. In 1988 alone, the Philippine Institute of Volcanology and Seismology tallied at least 5 earthquakes a day, countless active faults, and 20 active volcanoes.

The country might have entered a new political era in the mid-1980s, but it seemed an unlucky time, as a relentless series of severe disasters shook the archipelago. In 1990, an earthquake devastated northern Luzon, damaging major roads and nearly isolating the city of Baguio. The following year, the cataclysmic eruption of Mount Pinatubo in Central Luzon deposited millions of tons of ash that carpeted once-rich rice fields and filled riverbeds to the brim.

While staggeringly destructive, these disasters took relatively few lives. Typhoon Uring, which struck the Eastern Visayas in October 1991, was far crueler. Leyte and Samar are among the country's poorest provinces, and they are regularly battered by torrential rains. Typhoons typically last a few days before they clear, leaving water-soaked land covered with debris, but Uring's heavy rains continued more than a week. Leyte's capital, Ormoc, which is set along a broad bay on the island's western coast, was particularly hard hit. Though it is a small city, Ormoc sports grand buildings, which overlook a green, a low break-wall, and waves that, on most days, surge quietly onto the beach.

Early on November 5, after days of rain, Ormoc awakened to a catastrophic flood.[1] Most residents had no time to evacuate; survivors reported hearing a roar that sounded like thunder as the deluge hit neighboring homes. Muddy floodwaters rushed down the slopes of the foothills that rise steeply behind the city, destroying dozens of bridges and sweeping away homes along the embankments and in Isla Verde, an islet in the middle of the Anilao River packed tightly with the dwellings of poor families. As the mudslides reached the low-lying areas, they thundered through the streets and into the sea. In parts of the city, the deluge quickly reached 20 feet. According to early reports, as many as 8,000 people died in minutes, in less time, an Ormoc resident remembered later, than it took her morning pot of rice to cook.

The sheer number of dead overwhelmed Ormoc. In this predominantly Catholic country, wakes are important, week-long events where people pay regards to the dead and console the living. After Uring, there was little time for ritual. The bodies, which would decompose rapidly in the tropical heat, were bulldozed into piles and trucked to mass graves in the cemetery. For days, corpses washed up on the shores of Leyte, neighboring Cebu, and even Bohol to the south. More than 2,000 people, many of them young children, were carried out to sea by the floods and never found.

When I arrived four years later, Ormoc was a quiet city that on the surface showed little evidence of the flood. In the evening, children raced across the green. Couples and clusters of men perched on the breakwall, playing guitars and radios. In the heart of the city, shattered neighborhoods had been rebuilt.

On a closer look, signs of the tragedy were obvious. Used European and American clothes hung in the market, remnants of the aid that poured in and the volunteers who came from throughout the Philippines to help clean up the city. At the cemetery's entrance, a statue commemorated the dead, a large stone block topped with black wires stretching upward like twisted limbs. New grave markers lined the main path, and behind the barred windows of the mausoleums, from long rows of photographs wrapped in plastic, looked out faces of those the floodwaters had claimed.

No one readily forgets a disaster of such magnitude. Even in Cebu and Manila, people would mention how they had gone to Ormoc to help. In Ormoc itself, which then had a population of about 140,000, every family had been touched. People would point out which neighborhoods lacked running water for a year after the flood or recall that the mounds of bodies had been very high. Those who had buried bodies would speak reluctantly, if at all, about what they had smelled or seen or done. Simply saying, "I was there," or "I drove one of the trucks," implied enough. José (Pepe) Alfaro, the gregarious chair of Ormoc's *sangguniang panlungsód*, city council, routinely took visitors to tour the cemetery. He admitted quietly that the sound of running water still gave him goose bumps.

A native of Ormoc, Pepe was the head of the environment committee of the *sangguniang panlungsód* and the founder of a local environmental group, S.O.S. Earth. A spirited man in his fifties with thinning hair, Pepe was entertaining, even when talking about Uring. Without prompting, he repeated some of the black humor that the typhoon had inspired. In Leyte and Cebu, he said, for months no one would buy fish from Ormoc Bay, afraid of cutting through the scales and finding a finger or toe or just a piece of button lodged in the belly. The flood even had a few bright sides, he quipped. It made Ormoc famous, and it attracted new investors. From an environmental standpoint, it also brought a benefit. Ormoc Bay had been overfished, and the reluctance of people to eat its harvest for six months gave the schools of *bolinao*, *lapu-lapu*, and *tanguinge* a chance to recover.

While Pepe tried to make light of Uring's aftermath, he did not joke about its causes. The other disasters may have been natural, but Uring had a human cause: the clear-cutting of the once-forested hills behind the city. There had been earlier floods, he said, and in the years before the typhoon hit he had been expecting another inundation, but not one so severe. Although its costs had been high, he was not sure that the lessons of the flood had been learned, in Ormoc or elsewhere. Immediately afterward, "everyone was an environmental-ist," and throughout the country, people remembered what happened at Ormoc, he said. But it could happen again and probably would, there or elsewhere. Isla Verde remained dangerous, but squatters again lived there; and he was sure, considering the causes of the flood, that not enough had changed.

. . .

Philippine culture is often described as having been honed in an archipelago of small islands where living close together teaches the importance of being a good neighbor. Whether explaining their culture to a visiting writer or trying to understand it themselves, Filipinos mention some of the same deep-seated values: *Pakikisama*—the ability to get along with others. *Hiyâ*—the shame that people feel when they do something considered wrong or obtrusive. *Utang na loób*—literally a debt of the heart, the belief that some favors or acts of gener-osity are so significant that they could never be repaid. Even in the modern heart of Manila, despite the proliferation of text-based cell phones and beep-ers and the constant presence of Taglish—the colorful fusion of Tagalog and English—people still place a premium on face-to-face contact. The smoothness of social interactions, even subtle ones, matters deeply. *Tsismis*, the casual talk about people and events that translates imperfectly into English as *gossip*, offers a way to examine how people behave and get along. *Tsismis* can be a respectable pastime. It links the past with the present, binding communities and joining people who have been apart.

At first, explanations of how Typhoon Uring could have been so destructive had the spontaneity and unreliability of *tsismis*. Remembering the thunder they had heard before the floods hit, people in Ormoc speculated that a high tide had cascaded over the breakwall and into the city. They spoke of a *buhawi*, a waterspout or tornado, although the rubble had not piled into the telltale twisting path usually left by a *buhawi*. Tales circulated of trees and logs, transformed by floodwaters into powerful battering rams, that splintered bridges and buildings. People thought that, high in the hills, the long, green-blue Lake Danao had overflowed, spilling its cool water onto the city.

It soon became clear that nothing so exotic was to blame. Floods and mudslides had become common in logged-over areas; those in Ormoc just happened on a larger scale, in part because of the geography. Like many Philippine cities, Ormoc lies on a narrow coastal plain between the mountains and the sea. That shelf of land also serves as the floodplain of two rivers that converge above the city. The Anilao and Malbasag begin as short tributaries that meander down the steep foothills in a pattern called dendritic, after the fine branches of the head of a nerve cell. At some of the rivers' bends, small dams and debris often block the water's flow. During the dry season, the rivers might run just knee high in their deep channels, which often fill during the rainy season. During the colossal rains of Typhoon Uring, the banks' porous, volcanic soil grew waterlogged and unstable. Water built up behind the small dams and around the narrow turns and then broke through, cascading down the hills and into the city that lay in the way.

Disasters of any size elicit "if only's." If only debris and garbage had not blocked the river channels. If only people had abandoned their homes sooner. If only the children could have climbed farther—to the roofs and up the trees—and held on tighter. Floods on Ormoc's hills years earlier had not been as devastating, though, and no one could have imagined that the water would rise so high. As more came to be known about the causes of the tragedy, it turned out that the real regrets at Ormoc—the real "if only's"—had to do with the way that the watershed had been managed for decades before Uring struck.

Five wealthy families owned much of Ormoc's 11,000-acre watershed, the drainage area for rainfall and underground streams. Key political players, these families wielded enormous local power, and they typically had made their fortunes from their land.

The Larrazabals, who owned a large hotel on the waterfront, the Don Felipe, were among what some locals called the "super rich." The clan patriarch was Dudy's grandfather, Felipe, who arrived from Spain around the turn of the century. Since then, the Larrazabals had made money from logging and cattle ranching, real estate and sugar plantations. Two of Felipe's sons served as mayors of Ormoc, one for two decades during the Marcos era. One grandson was a member of the city council. A granddaughter, Victoria Locsin, had been elected mayor in 1987; she was voted out of office in 1992, in part because of discontent with how she handled the disaster. Her husband, who came from a prominent family of sugar planters from nearby Iloilo, had represented the province of Leyte in Congress.

In the 1950s, the logging families had begun clear-cutting huge parcels of the watershed and planting them with sugar and coconut, then lucrative exports on the world market. In time, the plantations covered the foothills, extending up to the riverbanks. They also displaced poor families who, if they could not work as field laborers, became squatters in the city or moved up the mountain seeking land to clear and farm. Commercial logging in the watershed was banned in 1967, but landless peasants and small-time bootleg loggers continued to cut illegally. By the late 1980s, they had pared away many of the trees left by the larger operations. Pepe ruefully described the hills with his characteristic humor: "If there were trees, they were as thin as the hair on my head."

According to foresters' conventional wisdom, to preserve the soil and keep the water cycle stable, trees should cover more than half of the area of the Philippines' small, mountainous islands. With Ormoc's watershed mostly planted with sugarcane and coconut, trees covered at best 10 percent of the land, not nearly enough to protect the soil and buffer the island from the ferocity of typhoons such as Uring.

After the initial, bewildered *tsismis*, as people realized that clearing the hills had made them vulnerable to mudslides, they first blamed the logging by peasants and bootleg loggers. Leyte governor Leopoldo Patilla called the tragedy a "man-made disaster caused by illegal logging in the mountains of the Visayas." Ormoc mayor Victoria Locsin cited "clear evidence of illegal logging operations." Before a week had passed, Fulgencio (Jun) Factoran, then secretary of the Department of Environment and Natural Resources (DENR), vowed to "expose illegal loggers and their political patrons." Newspapers carried daily reports of rampant unsanctioned logging in Leyte and elsewhere, and politicians quickly stood up to distance themselves from it. For a short time, the name *Ormoc* itself was synonymous with illegal logging.

While small-scale, illegal cutting contributed to the deforestation, it was largely the legal uses of the Ormoc watershed that had brought so much ruin.

For decades, government policies had favored commercial logging and supported the conversion of forestland for agriculture, despite the known ecological risks. It seemed too abstract, said Pepe, to blame the flood on the local elite, who had covered the watershed with plantations, or on the national policies that supported their actions. Instead, it was easier to blame illegal logging. "You can't say a flood like that was caused by improper land use," he said. "Who would care about that?"

The story of Ormoc is the story of uplands and lowlands around the country. It is just one example of the far-reaching consequences of largely unregulated logging in public forests, the conversion of forests to plantations, poor watershed management, and the near feudal control of land. Despite a facade of environmental laws, protected areas, and reforestation projects, in less than half a century about one-quarter of the Philippine forestlands were cleared, including those safeguarding watersheds. As loggers opened up new areas, landless migrants followed their roads into the forests and cleared land to farm, completing the deforestation. The forests' loss brought erosion, mudslides, flash floods, and drought. This ecological havoc further impoverished rural areas, helping deepen the economic decline.

Before Ormoc, floods and mudslides had caused only a few dozen or perhaps a few hundred deaths at a time. As news of the disaster reverberated from Luzon to Mindanao, people recognized that clearing the forests had transformed the archipelago. It was no longer possible to pretend that the remaining forests were being guarded as a perpetual natural resource or that denuded forestlands were being replanted. After the thousands of deaths, it was also harder to dismiss environmentalism as a reflection of western values. A young and tentative environmental movement had begun, and Ormoc gave it momentum. The city's name became shorthand for the flood and an emblem of what the enduring misuse of natural resources had meant to the rural provinces and the rest of the country.

⚡

Although there are few written records that describe the Philippine archipelago before it was colonized, it is commonly believed that when the Spaniards arrived in 1521, led by Ferdinand Magellan, forest enveloped much of the country. Because most of the forests were cut before they could be explored and studied, biologists consider them lost ecosystems. To visualize these extensive, lost ecosystems and understand their ecological roles, it helps to revisit the writings of western explorers in the 18th and 19th centuries who ventured into the forests spanning the equatorial tropics. They sent back stories, sometimes exaggerated, of forests that still capture the public imagination in more temperate climates.

According to their own accounts, the explorers were amazed to have discovered a natural world remarkably different from any they had known. Not only were the plants and animals far more varied, but to biologists trying to understand how life-forms evolved together over geologic time, they offered intricate ecological puzzles. The explorers' reports reveal their wonder. Alfred Russel Wallace, a contemporary of Charles Darwin, explored the Malay Archipelago in the mid-1800s. He included the Philippines as one of its "forest countries" whose islands were "almost always clothed with a forest vegetation from the level of the sea to the summits of the loftiest mountains."[2] The better-known Darwin wrote of the "delight" he felt as a 22-year-old first wandering in a Brazilian forest—"a deeper pleasure than he can ever hope to experience again."[3] Wallace, who eventually distinguished himself as both a scientist and social critic, was more circumspect. Although the forests possessed a "grandeur and sublimity altogether their own," for him the half-light also brought a "weird gloom and a solemn silence. . . . It is a world in which man seems an intruder, and where he feels overwhelmed by the contemplation of the ever-acting forces, which, from the simple elements of the atmosphere, build up the great mass of vegetation which overshadows, and almost seems to oppress the earth."[4]

The sheer diversity of the tropical forests awed biologists accustomed to the temperate zone's stands of beech, maple, pine, and oak. Wallace wrote: "Instead of endless repetitions of the same forms of trunk . . . the eye wanders from one tree to another and rarely detects two of the same species."[5] The stories and specimens that he and other travelers brought back profoundly influenced the development of biology. The work of Alexander von Humboldt in South America was critical to the study of ecology, and his observations of how vegetation changes with climate led to the new field of biogeography. Wallace and Darwin both relied on lessons learned in the tropics as they independently devised theories about natural selection. For "the student of nature," wrote Wallace, the tropics will "ever be of surpassing interest, whether for the variety of forms and structures which it presents, for the boundless energy with which the life of plants is therein manifested, or for the help which it gives us in our search after the laws which have determined the production of such infinitely varied organisms."[6]

While the explorers contributed to the growing field of biology, they also helped open up the forests. The European colonial powers were drawn to the tropics in part because of their rich resources, and the forests were particularly alluring. The trees, noted early explorers, would be a boon to the Spanish galleon trade between the Philippines and Mexico. "At least 10 ships can be built

shoes separate at the seams. In rural villages, dirt roads run with streams, then soften to thigh-deep mud. Lush foliage becomes even more lush, and rivers are replenished.

After the rains begin, the typhoons follow, great whirling winds that blow in from the Pacific, sending rain and debris in diagonal flight. Several typhoons may strike in a week, lighting the sky with swirling clouds and filling the air with the roar of wind and wind-driven rain. In cities, the storms flood low-lying neighborhoods, and in squatters' settlements they wrench roofs from walls. In the rural provinces, they batter the open land.

The forest's multiple layers can endure this onslaught. They buffer the high winds, deflect the hammering of rain, and shield the soil. They also modulate the water cycle. The rain drips through the many layers of leaves and runs down branches and trunks, soaking into the litter covering the forest floor. In what scientists call the sponge effect, the thick litter, soil, and plant roots absorb much of the rainfall, and over time, excess moisture seeps into underground streams and eventually into the rivers.

Clear-cutting the forests can bring both floods and droughts. Without the protection of the trees and surface litter, heavy rains cascade down the bare hillsides, washing away the soil, particularly the deep layer that acted as a sponge. As the rivers swell with tons of runoff, they scour their banks and overflow. The load of eroded soil accumulates behind dams and in reservoirs, reducing their capacity and shortening their useful lives. Where the rivers reach the sea, the fine soil particles settle onto reefs, smothering the corals. Fierce storms, as in Ormoc, can trigger flash floods and mudslides of waterlogged soil. Without vegetation absorbing the water and soil releasing it slowly, the dry season brings droughts that threaten the water supplies of cities and towns.

Clear-cutting can also permanently extract many of an ecosystem's nutrients. When western explorers first encountered the tropical forests, they imagined that they were supported by immensely fertile soil. This might have been true in the temperate zone, where topsoils can be deep and rich and organic matter decays slowly. A toppled tree may take years to decompose, and dead leaves can last a winter intact beneath snow. When the tree and leaves do decay, their nutrients become part of the rich topsoil and are drawn up by plants' fine roots. In humid tropical forests, nutrients cycle far more swiftly. Thick litter constantly forms on the forest floor as carcasses, scat, leaves, fruit, seeds, and even tree limbs rain down from above. In the heat and humidity, the myriad insects, fungi, and bacteria rapidly decompose the litter, leaving only a thin layer of topsoil.

As nitrogen, phosphorus, calcium, and other elements are released during decomposition, plants can capture them directly with networks of shallow roots

New rice fields.
San Fernando,
Bukidnon.
BARBARA GOLDOFTAS

and through symbiotic relationships with algae, fungi, and bacteria. This quick, efficient cycling means that most of the nutrients in tropical forests are held not in the topsoil, but rather in the thick decomposing litter and the lofty plant life. Tropical plants have adapted to the poor soil, and some species only undertake such high-energy tasks as setting seed and growing new leaves every few years. However, clear-cutting a forest and burning its organic matter strips the area of most of its nutrients. While the underlying soil in the lowlands can be rich, in other ecosystems—especially the uplands with their high temperatures and mildly acidic rainwater—over millennia the soluble minerals have leached from the soil and deep bedrock. Rather than being fertile, as the early explorers expected, tropical soils can be functionally sterile. Logged-over land often cannot support crops for more than one or two seasons.[10]

❧

When I first heard hills and islands in the Philippines described as "denuded," as they so often are, I assumed it was an exaggeration. Where I have lived in the U.S. Midwest and New England, even after years of cultivation, forests can reclaim open land. In the Philippines, many clear-cut foothills in the Philippines are covered with brush or grasses such as *cogon*, a tall, coarse grass that flourishes in disturbed areas. In other places, the sequence of logging followed by planting depleted the soil. I have seen foothills both on the outskirts of cities and in far-flung areas that look as bare and worn as the land in ancient Greece that Plato described thousands of years ago: "What now remains of the formerly rich land," he wrote, "is like the skeleton of a sick man with all the fat and soft earth having wasted away and only the bare framework remaining."[11]

While the Greek landscape became beloved for its stark beauty, in the Philippines, deforested landscapes have meant lost land, crops, and income. The costs of unbridled logging have been borne by farmers unable to feed their families who have fled to Palawan and Mindanao in search of land to clear; by refugees in Manila, whose squatters' settlements contain at least one-third of the urban population; by young women working as prostitutes who often come from a few deforested, impoverished provinces. While they might not call themselves ecological refugees, many of them had seen their surroundings change within one generation and knew that ecologically their provinces had gone awry. When people would speak of the forest, they would speak first of what they had lost.

Environmentalism is often cast as a luxury, a hindrance to development. According to this view, only citizens in industrialized countries, their basic needs met, can afford clean and healthy surroundings. In the Philippines, some environmental pressure has come from the small elite and middle class and from western influences, but early resistance to logging arose among those who bore the brunt of its effects. In isolated villages, where communication and travel were difficult and the military presence strong, opposing logging was dangerous work. One of the places best known for its opposition to logging in the 1980s was the municipality of San Fernando, Bukidnon.[12] Just as Ormoc became a symbol of the costs of ecological decline, San Fernando stood for the courage necessary to reverse that decline. In 1991, I visited the town to learn about what difference the courage had made.

San Fernando lies on the eastern side of Bukidnon Province in northern Mindanao. The town was originally called Halapitan, and in the 18th century it was inhabited, although sparsely, by *lumad*, indigenous peoples in Mindanao. The *lumad* organized the land into large communal forests where they practiced *kaingin*, swidden or shifting agriculture. After Ferdinand Marcos came to power in 1965, the names of many places were changed to honor his family; the *poblasyon*, town center, remained Halapitan, but the municipality was renamed San Fernando, a Spanish diminutive of Saint Ferdinand. One of the poorest municipalities in Bukidnon, by the late 1980s San Fernando had about 40,000 people, three-fourths of them refugees who were either landless or fleeing military conflict. About 10,000 were *lumad*.

San Fernando was the last jeepney stop on the road from Valencia, a 20-mile trip that, because of the poor roads, could take several hours. It was one of those places where the drivers would sleep in the jeepney, waiting in the morning until it filled with passengers before returning to town. At the station in Valencia, the jeepney to San Fernando sat for several hours before leaving in early afternoon. Bundles packed the aisle, and at the back, a plank had been laid between the two seats to hold extra passengers. Outside of Valencia, we passed through flooded rice fields and occasional towns. Made of unfinished wood, the houses were far apart and small, and they stood as boxy shadows among the fields and trees. As the road grew rougher, it passed through more fields that led to gentle foothills that led to the distant mountains. The jeepney was still full when we reached San Fernando, and everyone knew each other. A woman a row away introduced herself as someone I was coming to see. How had she known who I was? "There was no one else you could be," she said.

Bukidnon means "land of the mountains," which it is. Its high plateaus are cut by canyons, and in its peaks, seven wide rivers begin as headwaters, broadening as they flow into the far reaches of five adjoining provinces. Logging started in the San Fernando hills in the mid-1960s, as migrants arrived from more crowded areas. In 1980, five companies were cutting there. As in Ormoc, once commercial loggers felled the large trees, migrants, *kaingiñeros*, and small-scale loggers followed. By the late 1980s, the province's forest coverage had dwindled to about 14 percent, and government officials were discussing restricting logging.[13] In San Fernando, about 148,000 acres, or more than 80 percent of the town's area, were classified as forestland. The logging companies "would go day and night," said Father Charles (Kaloy) Gervais, one of several priests assigned to San Fernando. "When they knew time was running out, they logged with a vengeance."

The town of Halapitan had been carved from the forest not long before, and it was lush, with coconut palms, trees, and tree ferns crowding the sides of the packed-dirt roads, pressing into the open space. The *poblasyon* then consisted of a double row of stores, simple wooden structures with concrete foundations and open, glassless windows. It also had a monastery and a convent, often the main institutions in rural areas. The convent had low ceilings and tiny, dark rooms, where its rotating group of four to five nuns lived, according to their vows, a life of poverty. Across the street, the monastery was a well-lit, spacious place with wooden floors and a wide veranda where the priests would sit after dinner telling jokes and stories.

As in Malabog at the edge of Davao, the environmental activism in San Fernando was supported by the local clergy. In the early 1980s, members of basic Christian communities organized an environmental group, Pagbugtaw sa Kamatuoran (PSK), which roughly translates as To Be Awakened to Truth. PSK first focused on opposing a dam proposed by the national power company for the nearby Pulangi River that would have flooded San Fernando and other parts of Bukidnon. The project ultimately was abandoned, and as the ecological costs of deforestation emerged, PSK began to work with Father Kaloy and Father Patrick Kelly, an outspoken priest assigned there in 1984, to halt the logging.

Five years after the end of the Marcos era, the Philippine environmental movement was relatively new, and it still bore the stamp of the previous political era. It included organizations spawned by the religious community and research institutions as well as groups that had begun as nature societies such as the former bird-watching club, Haribon Foundation. Other groups, like PSK, had grown out of the broad movement opposing the Marcos dictatorship: local people's organizations as well as nongovernmental organizations (NGOs) with national and international links. Until the end of martial law, many had considered the state of the natural environment far less important than the ongoing political, economic, and military crises. The newer grassroots groups reflected the seriousness of that era. They tended to link ecological decline with its economic and political causes, emphasizing rural development as a way to limit the misuse of natural resources. The large national NGOs were based in Manila, but they ran projects in the provinces, where they worked with local people's organizations.

At this time, the environmental groups, like other parts of the civil society, still had an uneasy relationship with the government and military, which they had opposed under martial law. Their role then had been to provide services, especially to the poor, that the government could not or did not provide. Only gradually, sometimes begrudgingly, have the former adversaries—government and grassroots groups—changed their views of each other. Far more than the young people coming of age under democracy, those who had been active in the anti-Marcos movement usually remained suspicious of the government and military. Similarly, while older government officials might dismiss NGO workers as activists and rebels, younger ones might be open to working directly with the NGOs—or prefer doing so, if they had come from that sector. The influx of international funding and the growing number of NGOs also enhanced their influence. As national policies began to address the need for environmentally sustainable development and government officials pushed for reform within their agencies, NGOs and governments increasingly worked together.

. . .

The environmental organization in San Fernando had tried to make the transition from opposing to collaborating with the government. As its mission changed, its name also evolved, with typical Filipino attention to the nuances of language: PSK became KPPSK, which stood for Kapunungan sa Pagpanalipod ug Pagpalambo sa Kinaiyahan—Organization for the Protection and Development of the Environment. From 1987 to 1989, the group had held a series of protests that became more ambitious and effective over time. KPPSK first blocked the local road to logging trucks and then held a hunger strike at the DENR office in Manila. Their efforts drew national attention, and late in 1989, DENR secretary Jun Factoran agreed to ban logging in Bukidnon and set up a reforestation project. KPPSK provided staff, and by my visit, that was its only project.

I spent several days conducting interviews and at dawn one day joined KPPSK members walking to a reforestation site in the village of Kawayan several hours away. The group included Aming, a 28-year-old former Maryknoll missionary; Eskoy, a 35-year-old farmer; and Ruth (Ching) Esquillo, a 23-year-old graduate student from Ateneo de Manila University who was in her final weeks of researching the community's efforts at forest protection. They were spirited and warm; it was hard to imagine any of them throwing themselves in front of logging trucks.

Not far from the *poblasyon*, just before we entered a newly cleared valley, we reached the Tigwa River. Dry rocks lined the sides of the wide riverbed, and the water flowing through the center was shallow. We took off our shoes, rolled up our pants to the knees, and waded across. Eskoy, the farmer, had come from the Visayas in 1973, and when he first lived in San Fernando, the valley was forested and farmers still planted local varieties of unirrigated, upland rice. A small man with a narrow face and a quick, broad smile, Eskoy had seen the town transformed. "The first thing I saw [when I arrived] was the Tigwa River, which was very still and five meters wide," he said. "It was beautiful here. The water was clear, and there were many fish. San Fernando was forested then. Where we live now was forest. Where we lived then is now the town center. We grew upland rice, and we farmed in the spaces between the trees."

As the foothills were cleared, their soil eroded, the river became muddy with silt, and the fish died. Flash floods soon began to follow storms, he said.

"The first big flood came on February 24, 1982," the same day, he remembered, that his grandfather died. "Unmilled timber piled near the river washed away, scraping the sides of the river and felling trees. People were killed; *carabao* were killed. After that, every time there were heavy rains it flooded." The floodwaters would wash away fields along the riverbanks. One woman's family had owned about 10 acres of rice fields planted along a bend in the Tigwa. Only about two and a half acres remained—she waved an arm in that direction—near where we had crossed.

On the other side of the river, new rice fields covered the valley. Their raised mud edges had been laid so recently that they had not yet been trodden into an even walkway, and the flooded fields of young, light green plants were studded with charred stumps. This was all that remained of the forest: flanged trunks chopped off at angles, straight stumps that ended abruptly, chunks of toppled trees that burned where they had fallen. Several tall, narrow trees still stood: a decapitated palm, a fruit-laden palm, and oddly sized trees that alone had survived from the forest that had once pressed in around them and given them shape. I could barely imagine the creation of these fields: the downing of the massive trees, the hauling of the logs, the flight of the birds and mammals, the inferno that cleared the land and incinerated all life, the smoke and ash, the first blackened plowing of the soil.

Farther on, we walked through other fields filled with grasses or skinny, leafless trees that were dead but not charred. The narrow paths we followed led into the foothills, where the logging had been less tidy and thorough, fields had not yet been plowed, and brush and branchless trunks covered the land. The vegetation there was patchy and ecologically chaotic. At eye level, the three-foot layer of litter and soil was eroding visibly, falling away in chunks, crumbling like an abandoned stone wall.

In the late 1980s, after Marcos had been deposed, only two companies were still logging in San Fernando. Both had strong ties with politicians. The Caridad C. Almendras Logging Enterprises (CCALE) was named after the wife of former Senator Alejandro Almendras, a close supporter of Marcos, and El Labrador Lumber Company reportedly was owned by Congressman Victorio Chavez. At one point the logging concessions for these two companies in several provinces in northern Mindanao covered more than 225,000 acres. They also logged in San Fernando.

As the logging extended deeper into the uplands, its effects had become more pronounced. The deforestation seemed to have diminished the rain and lengthened the droughts. There was less water to grow rice, and farmers were losing their land to flooding. During the Easter Mass in 1987, Fathers Pat Kelly and Kaloy Gervais called for the community to do what the government would not—protect the remaining forests. "We have seen how the forests are being destroyed," the priests said during the sermon. "A few people, who are behind such destruction, know what will ultimately happen. But by that time, they won't be here any more to suffer the consequences. . . . Since those who are responsible for protecting the forests have not done their job, this responsibility rests on you, the people."[14]

A few days after Easter, members of KPPSK sent DENR secretary Jun Factoran a petition signed by 900 local residents asking him to cancel the licenses of logging companies operating in San Fernando. They sent copies of the letter to the local congresswoman, the district forester, the provincial military commander, and even President Corazon Aquino. When they received no response, they held a series of protests, using tactics honed during the martial law years. At the first protest, protesters claimed that the logging companies were cutting undersized trees, destroying roads and bridges, and not replanting logged-over areas. Early in the morning, people from the immediate area and distant *barrios* gathered in front of the municipal hall where trucks loaded with logs would pass. They barricaded the road with wood, logs, an abandoned culvert, and their own bodies. Signs they held told the drivers in both English and the local dialect to stop because "our irrigation has less water now!" and "the ecological balance is gone!" "Let us open our eyes," said another sign; "our forests are gone!"

Camping out in tents, day after day the men, women, and children covered the road, sometimes hundreds at a time, and turned back the logging trucks. Although the protesters had a permit from the local government, on the twelfth day, the loggers coaxed a restraining order from the court in the provincial capital, Malaybalay. The response was quick. The military, police, local paramilitary groups, and company goons violently dispersed the barricade, beating people with long sticks. Despite the violence, the picket continued by the side of the road, and after several weeks, the DENR agreed to temporarily suspend CCALE's logging permit.

No one in San Fernando was likely to forget the conflict. The area was part of the government's rural counterinsurgency campaign, which offered free health care and the alleged protection of the military to win back the "hearts and minds" of the people. Rather than making a place safer, the military detachments sparked violence. In San Fernando, the very armed units that had broken up the barricade were aligned with the loggers. As was true all along the forest frontier, opposing the loggers meant opposing the military.[15]

The local people saw no other option. The costs of ecological decline are often tallied with an eye to the future. Scientists argue, with growing concern, that deforestation and the greenhouse effect will raise the earth's temperature a few degrees, bringing global changes in infectious diseases and agriculture. The vast expanses of tropical forests, the argument goes, should be preserved in part to help regulate the worldwide climate. In the Philippines, concern about the climate has also focused on immediate costs—the chaos that ruined land and created ecological refugees.

In San Fernando, where the inhabitants intimately knew the land's daily and seasonal rhythms, the transformation had been quick and disorienting. One of the sisters at the convent, Sister Victoria Astorga, was assigned there in 1989. When she arrived, she said, she was "surprised to see that there were no trees. The mountains were very bare. The trees were only on the mountains far away." By that time, she said, "people could already feel the effects of the deforestation. Their fields were dry. The weather had changed—we couldn't feel when the rains would start." She and others described how, when they lived near the forest or under the umbrella-like reach of its branches, the shaded air was cooler; great mists, also cool, would rise in the morning, evaporating in tendrils under the sun. Since then, it had become hotter. The periodic droughts were longer, and the rains had become less predictable. When they did come, they were shorter and lighter, and they brought flash floods.

In the Easter sermon that inspired the first picket, Fathers Pat Kelly and Kaloy Gervais said that "forests attract the rains."[16] This simple explanation oddly captures the connection between trees and rainfall. While forests slow the pace at which rainfall seeps into the rivers, the trees also contribute to the water cycle, drawing up water in huge quantities and releasing it through pores in their leaves—a process called transpiration. In a large, dense forest, this water adds up, accounting for a significant portion of all rainfall. The Philippine forests were cut before the phenomenon could be examined there, but studies in the inland Amazon show that only about half of the rainfall that drenches the forests originates as water vapor blown in from clouds over the Atlantic Ocean. The other half is water that the forests recycle through transpiration, water that is lost, like the nutrients in the plant matter, when the forests are cleared.

As the logging continued, KPPSK took their protests to the regional and national levels, where the decisions that could bring local change would be made. In November 1988, hundreds of people from eight towns held a second picket several hours away in Malaybalay. They stopped 18 logging trucks, but this time the drivers, angry at the prospect of lost wages, charged with their vehicles, injuring some picketers. The provincial DENR did not act, but within a few weeks, DENR secretary Factoran himself came and suspended El Labrador's permit and canceled CCALE's. That action, too, had limited effect on the logging, and nearly a year later, 13 parish members and Father Kelly traveled to Manila. In front of the DENR's sprawling main office, they held a hunger strike, called a "Fast for Our Forest," which drew national attention. Three Bukidnon congressmen joined in pressing for a total logging ban in the province, and President Corazon Aquino again offered support. After 10 days, Secretary Factoran agreed to a 25-year logging ban, assigned 20 forest guards to San Fernando, and promised funds to replant almost 3,000 acres.

The reforestation program should not have been necessary. By law, the licensing agreements that allowed logging companies to cut also required them to replant as a way to ensure the forests' regrowth. Accustomed to little regulation, few companies followed the rule. Even when international lenders began funding reforestation projects on a large scale in the Philippines, relatively little land was successfully replanted. Much of the funding—hundreds of millions of dollars—was lost to corruption. "The company nurseries are just for show," said Aming, the former missionary. "When officials from DENR come, they're shown the nurseries, they're given *lechon*, and they never proceed to the areas to see what's really happened. They leave with envelopes in their pockets"— with bribes.

The replanting that was done was too often ineffective. The projects typically were badly planned and had inadequate resources. Funding from Manila could take a long time to reach rural provinces. The staff often were poorly trained; they might, for example, have the skills to plant the seedlings but not necessarily know how to maintain them. A lack of community support further hobbled the projects. Sometimes local people, including those who were involved in the replanting, torched or sabotaged reforested areas in other ways, hoping that if they could be paid once to plant trees, they could be paid to plant them again.

The replanting project in San Fernando offers a cautionary tale of the difficulties that local environmental projects can face. I visited during the second year of a three-year project. Partially funded by the Asian Development Bank, the project was intended to safeguard sections of the nearby mountain range and the Tigwa River's watershed in San Fernando. My early morning walk with the KPPSK members took us several hours through the valley and foothills before we reached Kawayan, a name that means "bamboo." I could not recognize the area as reforested. The gentle hills were scrubby, with the mess of bushes and grasses lacking any recognizable ecological pattern that often marks logged-over areas. The area was hot and sunbaked. My hosts pointed out recently planted seedlings, scrawny plants that, because of the overgrown vegetation around them, were easy to overlook. Most of the seedlings were just knee high, and many were dead. The nursery itself, which did not even have a shaded area to protect growing plants from the sun, was in shambles. It looked abandoned, and dead seedlings in plugs of potting soil lay in heaps on the ground.

According to Eskoy, the project had faced many obstacles, and the seedlings brought to the nursery were "not used much." With no vehicle, it was hard to transport fertilizer, so seedlings were often planted without the nutrients that they needed to survive in the poor soil. There also was not an adequate supply of water; the source had "dried up," he said. Funds from the DENR had been discontinued, and only 3 of the original 25 staff were still employed. Workers had stolen some of the materials—including the corrugated metal for constructing a flume to the nursery. "I'm very disappointed" in what was accomplished, said Eskoy. A woman working at the nursery insisted that they had planted more than 50,000 seedlings. KPPSK had monitored the operation, and Eskoy disagreed. The "records of seedlings planted were based on the number of seedlings sent from the [main] nursery," he said, and not on the numbers actually planted. Only about half of the seedlings planted had survived. Furthermore, after seedlings died, he said, the nursery workers "burned the bags they're in so when our monitoring team comes, the area will be clean. Project reports to the DENR say that everything is going fine. It's not."

KPPSK also criticized the project for planting species typically used in commercial plantations, such as *Acacia*, eucalyptus, *Falcata*, *Gmelina*, and mahogany. Unlike the native trees, these so-called fast-growing species mature so rapidly that they can be harvested within a decade or two. The use of commercial trees implied either a more complex system of watershed management than simple replanting or a lack of forethought. "This is an important watershed and needs to be protected," said Aming, "but the DENR is encouraging people to plant trees that can be harvested later. The trees are growing fast, which is good, but in 10 or 15 years, they'll be a temptation. People will want to [use them to pay to] educate their children. The next generation will fight each other again."

Even years after the protests, San Fernando remained an important symbol to the country's growing antilogging movement. People who had never been to Bukidnon knew how the farmers had linked arms across the road and halted the logging trucks. All along the forest frontier, those who out of necessity had become environmental activists spoke of San Fernando: in the westernmost province of Palawan, where trees were still burning; in the foothills of Davao in southern Mindanao, where remote parishes were just starting to oppose commercial logging; in a militarized village in northern Luzon, where a young minister outspoken about logging slept not in his bed, but on the floor of a back room where gunmen would be less likely to fire. At a time when local successes drew inspiration from each other, San Fernando offered hope.

In the town itself, success proved harder to define. The courage shown there may have given courage to others throughout the country, but in San Fernando itself, the logging continued illegally, like a terrible, destructive refrain. As elsewhere, the logging ban had served to increase illegal cutting, and no one could stop it. My hosts did not actually call their victory short-lived, but they did say that the reforestation project had failed and that, without the support of local officials, monitoring the logging was difficult. Like the source of water in Kawayan, the DENR funds had dried up, and while the forest guards continued their work for a few months without pay, they eventually had to stop.

KPPSK continued to confront the local powers and its own limits. The small organization had faced growing pains in trying to establish itself more formally. While the activism had drawn attention at a high government level, the group found other responsibilities difficult to handle. Ching Esquillo, who researched the KPPSK for her thesis, wrote that the "community spirit which inspired and sustained the successful pickets and fast" was not enough to sustain the reforestation project, which required more managerial and technical skills than the organization possessed. Tensions developed among the core members, and Father Kelly, whose leadership had been crucial, was reassigned to a different parish.[17]

KPPSK also did not have a smooth working relationship with the DENR. Said one member, "The [provincial] DENR is not following through in its responsibilities. We've given them all the paperwork, but they did nothing. The people laugh at us. We're frustrated with our inability to really protect the environment." Although NGO workers came to play a key role in joint efforts with government officials, they could not easily see each other as allies. "When the government came in to do the reforestation, it was very difficult at first," explained Ching in an interview in the late 1990s. "People see the government as corrupt, so there was tremendous distrust." The government workers in turn viewed local organizations as threats and believed that the KPPSK members were activists and that their involvement in the reforestation was "meddling."[18] As was happening around the country, the two groups were just learning to work together.

Together with the tragedy at Ormoc, the San Fernando protests helped galvanize environmental efforts in the Philippines, governmental and NGO alike. They showed that the ecological decline was widespread, that it was not inevitable, and that it brought considerable economic costs. Both taught the lesson that ecological destruction has roots in poverty as well as power and that without farmland or jobs, the poor, although less destructive overall than commercial operations, would cut trees and clear land that otherwise might have regrown. The nascent environmental movement brought changes in government policies, challenges to corruption, and curbs on the logging industry. At the local level, where commercial logging continued, it also brought bolder opposition.

In places such as San Fernando, though, change comes slowly. Trucks carrying fat logs passed on the road while I was there, and the mayor, who had been the police chief during the Marcos era, did not apologize for the fresh lumber leaning behind his home. A huge public gymnasium was being built—illegally—of wood. "Businesses and government employees and others who can buy lumber feel free to construct houses of wood. We have no ability to apprehend them," said a KPPSK member. "Lumber is delivered to the sawmills at night. The mayor, who is supposed to lead, tolerates the activities. He allowed contractors to build the gym using confiscated lumber. It was under the custody of the police, and it disappeared."

Despite the national recognition and international funding that KPPSK had received, it had limited clout, and in this small, lush town where everyone knew everyone, its supporters still lived in fear. The nuns would not cross the road alone at night. KPPSK members still drew threats. "We have been told," two women said, "to refrain from walking by ourselves." Whenever they left the town center, they told me, "we walk in groups of three or more. We never walk alone."

3 | To Get a Feeling for a Place: *Naninibago*

As a young priest, Nerilito Dazo Satur made his mark by becoming the 31st forest guard in the Philippines to lose his life. A well-loved priest, Father Neri had served a cluster of remote parishes in Valencia, several hours from San Fernando. Despite a ban on felling trees in Bukidnon, cutting persisted, and Valencia alone had three sawmills where logs were delivered at night. His work was dangerous, but Father Neri had been a determined opponent of the illegal logging who had single-handedly confiscated many truckloads of valuable logs. His vigilance angered the powerful local forces in the area, and in October 1991, as he returned from a religious fiesta with a young churchworker, the 30-year-old priest was gunned down on his motorcycle. Father Neri was not the first priest to be murdered in Mindanao or Bukidnon, nor would his death be the last in the antilogging movement.[1]

During its 300 years in the Philippines, Spain failed to fully colonize Mindanao. Its western provinces, especially parts of the Sulu Archipelago closest to Indonesia, were home to most of the country's *Moros* who, with little of the pressure from the Spaniards and the Catholic Church that had transformed much of the country, were able to keep their culture relatively intact. Historically, the greater the distance between a province and the capital, the less attention it received from national officials; economists still use distance from Manila to roughly measure access to government services. Mindanao's provinces, especially the Muslim ones, are among the country's least developed areas. Compared with the rest of the Philippines, the island has rougher roads, fewer hospitals and manufactured goods, and less information, and its official rates of literacy, life expectancy, and education have lagged. Even in 1991, fewer than 20 percent of families in the Sulu Archipelago lived in homes with electricity—less than half the average for all rural areas in the country. That same year, about half of Filipino families who reported having no access to running water—they used streams, rivers, or rainfall—lived in Mindanao.[2]

At the time that Father Neri was killed, from the vantage point of the capital, the sprawling island seemed a distant outpost. Flights to Mindanao from Manila's domestic airport were often delayed, the ferry there took more than a day, and the phone lines, where there were any, offered a tentative connection. Friends who had grown up there but moved to Manila to study or work or escape the divisions that had wracked the New People's Army reminisced about the freshness of the air in Mindanao and the beauty of its forested hills.

Throughout the 1990s, Mindanao remained a place of violence and conflict. *Moro* separatists continued their guerrilla wars for independence. The government-sanctioned paramilitary troops, called Civilian Armed Forces Geographical Units, or CAFGUs, ruled in the villages, where they served as the military's front line in its ongoing effort to quell the insurgency. In many rural areas, local warlords protected their modern-day fiefs with private armies. Gangs of kidnappers struck more frequently than elsewhere, often targeting foreigners. Independent fundamentalist vigilante groups, one of them ominously nicknamed *tadtad* (chop-chop), also terrorized local communities. From Manila, Mindanao seemed a lawless place with too many guns and unfettered local power.

Although the hub of central Bukidnon, Valencia then was a sleepy hub. A town of about 100,000 residents, Valencia had a market that drew farmers from outlying villages who would travel by jeepney for hours with produce to sell. The jeepney station was a square ringed with small eateries and vehicles waiting for passengers. Within view stood the monastery where Father Neri had lived and worked.

I arrived in Valencia shortly after the national newspapers reported Father Neri's death. In some places, local violence could remain invisible to outsiders; faces would shutter closed at direct questions, or people would explain vaguely, and with evident resignation, the difficulty of living near a military detachment or in an area of "conflict." In Valencia, the murder drew a response that was public and defiant. At eye level on the heavy, wooden doors of the monastery were tacked about 10 color photographs documenting Father Neri's murder. They showed his body lying in the road, the packed earth around him soaked with blood, his coffin, a funeral procession, flowers. They told the story that priests inside did not hesitate to relate.

Ordained just two years earlier, Father Neri had been assigned as assistant parish priest in Barangay Guinoyoran on the outskirts of Valencia. Known as a hideout for bandits, Guinoyoran endured an army detachment, aggressive CAFGUs, and members of the *tadtad* vigilante group. Over the previous three years, there had been 13 murders, most of them reportedly linked with at least one of these armed groups, and local people referred to that part of Guinoyoran as "the killing fields." Father Neri was a roving priest who serviced the remote areas, traveling from village to village on his Honda 250 motorcycle and saying mass in a different place every week. Outgoing and outspoken, he was apparently well liked, but he was known best as an energetic opponent of illegal logging and gambling.

Father Neri's public stance on logging was not unusual in Bukidnon or religious circles. His activism, like that of many others, had been spurred by a call from the religious leadership. In January 1988, the powerful Catholic Bishops Conference of the Philippines released a long and at times passionate pastoral letter on ecology that became influential, particularly in rural areas where communities were facing the effects of deforestation. Titled "What Is Happening to Our Beautiful Land," the pastoral letter warned about the "deep-seated crisis" that lay at the root of many of the country's economic and political problems.[3]

"One does not need to be an expert to see what is happening and to be profoundly troubled by it," wrote the bishops. They cited damage to the forests, rivers, land, and oceans. They also spoke scathingly about the social and economic consequences of ecological decline and the further upheaval that it could bring: "[F]or the vast majority of Filipinos, the scars on nature, which increasingly we see all around us, mean less nutritious food, poorer health and an uncertain future. This will inevitably lead to an increase in political and social unrest."

The pastoral letter also warned against complacency and a "fatalistic attitude." Recalling the courage of people in San Fernando and elsewhere who had "defended the remains of their forest with their own bodies," the bishops pressed for immediate action throughout the society. They urged each individual "not to remain silent when you see your environment being destroyed." The Catholic churches should redress their "slow [response] to the ecological crisis" and at every level of church organization create a "Care of the Earth Ministry" to focus on ecological concerns. The bishops also urged the government to work harder in its policies, programs, research, and enforcement, while pressing NGOs to serve as watchdogs to ensure that officials "do not renege on their commitment."

In response to the bishops' call, people all over the country formed ecology committees at their churches and tried to halt illegal logging—by monitoring cutting, intercepting shipments, and documenting violations and violence alike. They also worked as forest guards. In the late 1980s and early 1990s, the Department of Environment and Natural Resources (DENR) employed only one forest ranger per 7,000 acres.[4] To bolster enforcement, officials deputized civilians, giving them the authority to confiscate illegally cut lumber and otherwise help safeguard the forests. In remote communities, government rangers might be less effective than the local forest guards, who knew the forests and, probably, the illegal loggers. While the government employees tended to operate from a base in town, the forest guards were often willing to patrol at night deep in the mountains where they could hear the chain saws buzz and act quickly to apprehend loggers. As the forest guards joined the drawn-out struggle, their effectiveness turned them into targets.

Within the church hierarchy, reactions to the clergy's activism differed. In Malabog on the outskirts of Davao City, Father Max's predecessor had been reassigned because of his activism against logging. In Bukidnon, the bishop encouraged the clergy to join the battle to preserve the province's forest. In 1990, he and 45 priests, including Father Neri, were deputized by the DENR as forest guards. Father Neri took the role seriously; outside a local sawmill, he apprehended 14 shipments of rough-cut lumber. His efforts did not go unnoticed. "He was very bold, out front about it, very vocal, talked about it in his sermons," said a nun a few hours away in San Fernando. "He really did his job, did it too well. He got killed for it."

In mid-1991, Father Neri and two other priests began receiving death threats. Several weeks before his death, parishioners cautioned that for his own safety, he should not attend an upcoming fiesta. Heeding the warning, he stayed home. The next fiesta, a Eucharist celebration, was at Sitio Tambulan in Barangay Guinoyoran. This time he went, traveling with a young female churchworker, a university student who rode behind him on his motorcycle. At midday on their return trip, about seven miles from the center of Valencia, they were ambushed.

When I visited the monastery, a handful of priests and parishioners stood about, talking in low tones. They repeated the story told by the churchworker. Three men waiting with a rifle, a handgun, and a homemade shotgun shot Father Neri in the chest. As he lay pinned by his motorcycle, one of the men approached and smashed his head with the butt of the shotgun until the handle broke. The church worker had been shot in the leg, but the attackers told her to run. They had no grievance against her, several men explained to me, and there was a taboo against fighting girls. It would have brought *hiyâ*, shame, if they had killed her.

The parish priest, 34-year-old Father Loloy Sangelan, explained that he had assigned Father Neri to Guinoyoran. "It's very painful," he said. A padlock secured the narrow door to what had been the priest's room, a small space with little more than a window and a sink, a slim bed, and a bookcase. On the shelves lay Father Neri's motorcycle helmet, a pile of coins he collected at mass the morning he died, some clothes, his heavy boots, and a scrolled 30th birthday card inscribed with many good wishes. Father Neri used to wake Father Loloy for morning mass by knocking on the wall between their rooms. "We were here two years together," he said. "We shared our feelings about the parish. He was like a brother to me."

⚘

The Tagalog word *naninibago* means to get a sense of a new place. It is a verb built around the word *bago*—new. After I arrived in Manila, Danny, a young NGO worker from Mindanao who accompanied me on interviews, tried to acculturate me by teaching me this concept. He used the English translation. When I would ask questions about a place, Danny would quietly encourage me to try to find a way to answer them myself. "Get a feeling for the place," he would say. "Try to get a feeling for the place." As he shepherded me from office to office and provincial town to provincial town, he would point out nuances worth noticing: words, voices, body language, polite shyness that relaxed into warmth, the way that people glanced at each other, or away, after a question.

If he had been in Valencia, Danny would have noticed the tension there, the suspiciousness of strangers, the guardedness and hesitation that can suggest that somewhere nearby there are guns. I was the only foreigner on the streets, and eyes tracked my path. People would talk after I left, note where I had been and whom I had visited, even though other journalists had already passed through Valencia to write about the young priest.

It was a few weeks after Father Neri's funeral, and grief and chaos still filled the monastery. The mass in his memory had been the only mass held in the huge province that Sunday, and it was attended, the newspapers reported, by more than 60 bishops and priests, 100 religious, and 5,000 laypeople.[5] Quiet with grief, Father Loloy was also angry. The deanery would not assign anyone else to Guinoyoran, he said, but he *would* continue the work for which Father Neri had been killed. An arrest warrant had been issued for three men, including Catalino Cabison, a sergeant based at the Guinoyoran army detachment, but they had not surrendered.

The threats from Cabison, Father Loloy had heard, continued. "Before, Cabison said that he wanted to kill three priests. Now Father Neri is dead, and he said that if we don't stop, we'll be next." Father Loloy paused. "I do think that if we continue apprehending illegal loggers, the same thing will happen to us.

But if the people come to me . . ." He paused again. "I'm afraid," he said, "but it's part of my duty in this parish. If I don't do anything, the next generation will blame us."

After I left the monastery, I walked back through the town, accompanied by a young *novicio*, a priest in training. Father Loloy had directed him to guide me to a shed near the police headquarters where lumber that Father Neri had confiscated was stored. We first went to ask permission from the police.

In the rural provinces, permission often had to be asked. This was in part out of *galang*, respect for hierarchy and authority; before I could interview a government official, I would need to be introduced to the mayor or vice-mayor or immediate supervisor. I was also—and this was probably more for my hosts' safety than my own—often introduced to the police chief. Many rural areas were still militarized, and to enter, I needed permission from the officer in charge. Even to look at logs confiscated by a government-appointed forest guard, it was best to start by telling the local police. This caution had less to do with *galang* than with fear and practicality. The *novicio* knew who wielded power in Valencia, and he knew the value of logs, even though those we saw hardly seemed worth a life.

The headquarters turned out to be closed—it was a holiday, said one of a small group of men standing there. Apparently satisfied with the effort, the *novicio* led me to a nearby open shed with a corrugated metal roof under which the wood lay stacked. The knot of men watched us, but we did not look back. The trees had been small, and the logs were sawed into irregular lengths, each about as high as a child's knee. On the ends of some were painted numbers in white: 1, 6, 53, 77, 97. There were many dozens; Father Neri had worked hard. Some of the logs were not numbered. Some, said the *novicio*, were missing.

As we looked at the wood, a few more men gathered behind us. Armed but not uniformed, they stood casually. I knew without being told that they were a CAFGU. The *novicio* said nothing. Three little boys in shorts and rubber flip-flops clambered over the wood; they posed while I took photographs, using them for scale, because afterward it would be hard to remember the size of the logs. The *novicio* and I stood stiffly, the eyes of the CAFGU on our backs. When we walked away, the men followed us in an open truck down the hill to the monastery, holding their long guns.

◆ ◆ ◆

In 1898, when they defeated the Spanish fleet in Manila Bay, the Americans controlled just Manila and its surroundings. The Filipino nationalists, who had long been fighting for their independence, held the rest of the archipelago. They collaborated with the Americans against Spain, believing that doing so would bring them liberty. The Americans instead made them a colony, together with Cuba, Guam, and Puerto Rico, and emerged as a new global power. The Filipinos resisted, and the Philippine-American War, well remembered in the Philippines but nearly forgotten in U.S. history, lasted 3 years officially and 10 unofficially. Marked by violence against Filipino civilians, the war took hundreds of thousands of lives.[6] Although granted commonwealth status in 1934, the country did not gain full independence until after World War II.

The acquisition of the Philippines and the war that followed drew opposition in the United States. The Anti-Imperialist League, founded in 1898, held that countries should be able to govern themselves, as the United States had fought a revolution to do. Imperialism, stated the league, was by definition un-American. The U.S. senator and former secretary of the interior Carl Schurz described the Philippine-American War as "criminal aggression."[7] Mark Twain, who devoted some of his last years to the anti-imperialist cause, was another formidable spokesman against the war. "We have gone there to conquer, not to redeem," wrote Twain. "And so I am an anti-imperialist. I am opposed to having the eagle put its talons on any other land." As the war progressed, he condemned the brutalities against the Filipinos and wrote harshly about the confinement of civilians in concentration camps controlled by the U.S. military. The war itself, he wrote, betrayed the ideals of the United States.[8]

Those who supported American expansion justified its acquisition as "the white man's burden"—a phrase taken from a poem by the British poet Rudyard Kipling about the U.S. taking of the Philippines. Imperialists claimed it was their duty to bring civilization to people who could not govern themselves. They believed that the Philippines was also important for military and economic purposes.[9] The combination of the two left its mark. The Americans deeply influenced the development of the logging industry and the forest service and their approach to forest management and conservation.

One of the greatest champions of the U.S. role in the Philippines and the archipelago's commercial potential was Dean C. Worcester, who first went there as a biologist and explorer. As a student in Michigan, he had admired the dozens of unknown species of tropical birds collected in the archipelago by a zoologist, whom he joined on his next trip. Worcester took two long expeditions between 1887 and 1893, exploring isolated villages and collecting specimens. He ultimately lived nearly half his life in the Philippines; he was appointed to the first two Philippine Commissions, served as secretary of the interior for 13 years, and died a wealthy businessman in Manila at the age of 57.

Worcester contributed to the public debate over U.S. policy in the Philippines through speeches, articles, and two books that influenced American views of the new colony. Although he idealized the United States, he was contemptuous of Filipinos, whom he believed to be childlike, indolent savages who lacked the capacity to govern themselves. That view colored his administration of the colony and the opportunities that he provided for its economic and political development.[10]

Another advocate for U.S. control of the Philippines was Senator Alfred Beveridge who, after returning from a tour of the archipelago in 1900, exalted the new colony's virtues in a speech to the U.S. Senate. "[T]he Pacific is the ocean of the commerce of the future. Most future wars will be conflicts for commerce. The power that rules the Pacific, therefore, is the power that rules the world. And, with the Philippines, that power is and will forever be the American Republic."

Beveridge stressed the value of the potential markets in the Philippines and its strategic proximity to China—and those "illimitable markets." He also praised the benefits that the Philippines' ample natural resources could confer to the growing American power, especially at a time when the U.S. logging industry was peaking: "I have cruised more than 2,000 miles through the archipelago, every moment a surprise at its loveliness and wealth. I have ridden hundreds of miles on the islands, every foot of the way a revelation of vegetable and mineral riches. No land in America surpasses in fertility the plains and valleys of Luzon. . . . The wood of the Philippines can supply the furniture of the world for a century to come."[11]

Like Beveridge, Worcester appreciated the forests' commercial prospects. "Certainly no other country has a greater variety of beautiful and serviceable woods," he wrote. There was wood of various colors; wood so strong it was "almost indestructible even when buried in the earth." There was cane for wicker, the endlessly versatile bamboo, and trees that could supply "valuable gums . . . alcohol, tan barks, dyewoods, valuable vegetable oils or drugs."[12]

Although anti-imperialism failed to gain support in the United States, another idea—that natural resources should be conserved—garnered more attention. In the 1860s, forests still blanketed much of the western United States. People commonly believed that the nation's resources were inexhaustible; but smugglers were cavalierly sawing down the great old-growth trees, and concern grew over what the loss of the forests might mean—to the soil and water cycle, the logging industry, and the country's economy. Scientists, government officials, and politicians began to discuss how the forests could be maintained, a debate that eventually gave rise to a conservation movement. In the Philippines, Dean Worcester had recognized the value of the forests: "Surely the Philippine forests should be preserved, but not for their beauty alone! In them people have a permanent source of wealth."[13] This wealth ended up being temporary—partly because of the direction that the debate over conservation took in the United States.

. . .

The story of how the U.S. conservation movement developed and ultimately affected the Philippines can be told through some of its participants. An early voice was that of George Perkins Marsh, a scholar later appointed as the U.S. minister to Italy. Marsh came from Vermont, where the Green Mountains had been cleared of their forests. Involved in logging and selling lumber himself, Marsh had witnessed profound ecological changes brought by deforestation, changes "too striking," he declared in 1847, "to have escaped the attention of any observing person." Like Filipino farmers a century later, he observed droughts, floods, and depleted soils, and he feared that "the valleys of many of our streams will soon be converted from smiling meadows into broad wastes of shingle and gravel and pebbles, deserts in summer, and seas in autumn and spring." He had seen such "derangements," as he called them, in deforested parts of Europe, and he believed that "tree cover was a vital attribute of nature." In some European countries, laws already safeguarded watersheds and limited cutting, protections Marsh hoped to put in place in the United States.[14]

In 1864, Marsh published a book called *Man and Nature* to document the "ravages committed by men" and to give "the most important practical conclusions suggested by the history of man's efforts to replenish the earth and subdue it." There was not yet a body of scientific theory and evidence about what careless management of land might mean in the long term. "[S]ystematic observation . . . has barely begun, and the scattered data which have chanced to be recorded have never been collected," he wrote. Nonetheless, he strongly encouraged less "profligate waste." The word *ecology* would not be coined for a few more years; the phrase used instead to capture the natural world's interconnectedness was the *economy of nature*, which he invoked. "The world cannot afford to wait till the slow and sure progress of exact science has taught it a better economy."[15]

The battle over how to use, conserve, and ultimately manage the vast, government-owned U.S. forests—a battle that would last more than 40 years—had barely begun. One of those who found the case for conservation compelling was Franklin B. Hough. A physician, historian, and statistician, Hough had noticed, while scrutinizing forestry census data in the early 1870s, that timber production was falling in some areas but rising in others. Timber supplies, he realized, were being depleted, which raised the specter of shortages. The government had a duty to preserve the forests, he argued to the American Association for the Advancement of Science, and a few years later, he was appointed chief of the forest service. As reckless cutting continued in the timberlands, Hough, like Marsh, denounced the frontier mentality. Believing that private ownership alone could not ensure sound land use, he proposed a system of leasing timber privileges for the public forestlands. Modeled after a program that had successfully conserved the Canadian forests, it would allow loggers to cut for a fixed number of years but prohibit them from cutting younger trees.[16]

Hough fought for 10 years to create public forests and died without seeing his efforts succeed. His work, however, influenced Gifford Pinchot, often called the father of conservation in the United States. Pinchot organized the U.S. Forest Service in the early 1900s and, as its chief under President Theodore Roosevelt, worked hard to build forestry as a profession. He also oversaw the establishment of 63 million acres of forest reserves, later called national forests, which by 1907 had grown to 150 million acres.

A split had formed early on among conservationists and early environmentalists, the precursor of a division that endures today. Some, like Sierra Club founder John Muir, believed that nature, particularly wilderness, should be valued in its own right rather than just for what it might offer people. They also saw the natural world as a place for recreation and as a unique refuge, especially from a world that seemed to grow increasingly modern and separate from the natural one. Others regarded nature strictly as resources to be used—the "permanent source of wealth" that Worcester saw in the Philippine forests—and believed that these riches should be safeguarded for the future. Pinchot firmly held the latter view. He believed that nature should be managed, not preserved as wilderness, and he valued the forests for their timber and for the roles they played in protecting the soil and the water cycle. Although the extensive forest reserves were set aside to protect the trees and watersheds, mature and dead timber could be cut and sold. The "prime object of the forest reserves is use," declared the pocket-sized manual of rules and procedures issued to foresters that Pinchot pointedly named the *Use Book*.[17]

The mark Pinchot left on the U.S. forests is well known, but less remembered is a parallel system he set up in the Philippines. In 1900, one of the first major actions of the U.S. military government in the new colony was the creation of a forestry bureau (ostensibly, according to the United Nations Food and Agriculture Organization, to "harness the forest for military logistics").[18] Two years later, Pinchot visited the Philippines and spent six weeks with one of his protégés, Captain George Ahern, a forester who would become the Forestry Bureau's director. Together they toured the archipelago on a U.S. gunboat, inspected the forests, and drafted a forest law. The charter for the Forestry Bureau echoed that of the U.S. Forest Service: "The public forests and forest reserves of the Philippine Islands were to be held and administered for the protection of the public interest, the utility and safety of the forests, and the perpetuation thereof in productive condition by wise use."[19]

The system of government that the U.S. Congress approved for the Philippines in 1904 included the forestry law sketched out by Pinchot and Ahern, which would serve as the basis for future forest management. The law designated the timberlands as public property controlled by the Bureau of Forestry, which had the power to issue timber licenses. In 1910, a forestry school was set up at the University of the Philippines in Los Baños near Manila, and Ahern drew graduates from the new U.S. forestry schools to help plan a modern industry to make full use of the wealth of trees. At that time, the bureau oversaw nearly 50 million acres of forest, about two-thirds of the country's land.

The Spaniards, who had claimed the once communally owned forestlands as public property, cleared some forests for massive plantations and cattle ranges, including those covering the central plains of Luzon. They also used timber for furniture and cabinetry; the island of Cebu was largely denuded to build the Spanish galleons. During this era, the forest coverage declined from about 90 to about 70 percent of the country's total area. Known for neglecting the economic development of their colonies, the Spaniards did little to foster industries based on the forests, and they cut without replanting.

Ahern recruited U.S. lumbermen to invest, and his efforts helped launch the Philippine logging industry, which was the first and most technologically advanced in Southeast Asia. Because of exports to Europe and the United States, Central American mahogany was already in decline, and Ahern introduced dipterocarp wood to the U.S. market, calling it by the misnomer "Philippine mahogany." In 1904, he awarded the American W. P. Clark the country's first logging concession: 20 years' access to 115 square miles in northern Negros. A major manufacturer of sawmill equipment from Seattle, Washington, Clark shipped to the Philippines a replica of the most advanced mill in the U.S. Pacific Northwest. He set up the Insular Lumber Company in Negros, which became a leading lumber company that used modern locomotive and milling technology, eventually clearing the land on the island for sugar plantations.[20]

The Philippines also contributed to the growth of the U.S. timber industry by importing lumber for construction. Although their own wood industry was developing, it was oriented toward export, so the islands were also becoming a market for imported wood; Manila and other growing port cities were built from redwoods and Douglas firs shipped from Washington and Oregon.[21]

Like its U.S. counterpart, the new forestry bureau stressed technological modernization and the scientific management of forestlands as a way to sustain

ones had burned in an 1897 fire in Manila, and others were destroyed during World War II. Since then, the poor record-keeping reflected widespread corruption. Some documents simply disappeared, including the results of a key seven-year inventory begun in 1954. Other official figures proved "of dubious quality." There was, Kummer noted with evident frustration, "substantial circumstantial evidence to indicate deliberate manipulation or destruction of data by government officials. I personally have encountered all of these problems." Even the widely used rates for deforestation turned out to be inaccurate, and some significantly underestimated the problem, including government figures for most of the 1980s that were "so low as to have virtually no credibility." Overall, he concluded, the government data on both exports and forest cover were manipulated so as to mask the extent of the problem and "mislead the Filipino media and forestry community and foreign researchers."[27]

While the statistics masked the magnitude of the logging and exports, the warnings about overcutting too often focused on the small-scale loggers and also on *kaingin*, shifting agriculture. In indigenous cultures, *kaingin* was originally bound up with religion. The indigenous people did not own land, which they viewed as a divine gift, and they linked forests with specific tribal gods. A council of elders and a *datu* oversaw the forests and land; their careless or destructive use was believed to anger the deities. People asked permission to do *kaingin*—to make a clearing in the communal forest and plant crops. They took care not to damage adjacent land or trees, and to maintain the fertility of the soil, they moved—shifted—the plots regularly. Sacred areas and hunting grounds were excluded from cultivation.

As the forests were opened up, millions of migrants from the Visayas or elsewhere in Mindanao moved into the foothills. Most were fleeing poverty or military conflict. By the early 1990s, as many as 18 million people were believed to be living in the uplands, on earth too steep and too poor to support them.[28] The indigenous people, confined to smaller and smaller areas, abandoned and then forgot many of the traditional practices. The migrants from the lowlands—called refugees or settlers—also burned forest to clear it, but they knew nothing about the traditional systems of cultivation. Rather than rotating their tilled plots, they would farm in one place until they exhausted the soil. Nonetheless, their fires were indistinguishable from those of the *lumad*, and all farmers clearing forestland came to be called *kaingiñeros*.

As early as 1905, a report identified *kaingin* and illegal cutting as the primary cause of the forests' destruction, as local officials do still.[29] Because the small-scale loggers and farmers did the final clearing, they often were wholly blamed for the forests' loss. The commercial loggers typically felled the larger trees, and they claimed that this selective cutting left the forests relatively unscathed. Until the loggers came, though, the forests remained largely impenetrable, except along their edges. Once they opened up the forests by building logging roads and felling the large trees, the forests became accessible to small loggers and farmers. Danilo Balete, a biologist with the Haribon Foundation, said, "Once you build roads and clear the forest, it's so easy for migrants to go and finish up the job. That's one reason why people blame *kaingin*—because that's what they see. They don't consider how it all started."[30] Blaming the landless farmers laboriously clearing small plots of land also deflected attention from the deeper causes of the deforestation, particularly the roles played by the logging companies and government.

◆ ◆ ◆

Historian David Joel Steinberg has called the Philippine timber industry "one of the most corrupt [industries] in the country," describing a pattern of "kickbacks, dummy corporations, bribery, and violent crime" that make it a "national disgrace."[31] By law, companies had to replant a hectare of land for every hectare logged, but despite ongoing and often official declarations about reforestation, the companies did little replanting. While the logging was extremely profitable, it did not help develop the rural areas. Although the export trade provided many thousands of jobs in the provinces—even as production slowed and foreign markets flagged—the jobs paid relatively little and did not build the local economies.

A generally weak government with close ties with the timber industry further fueled the logging. In the past, working in government brought not a means to offer service, but rather access to resources, a way that Filipinos might enrich themselves and their families. Government officials could provide gifts, money, and jobs, either in exchange for votes or as a way to repay an *utang na loób*, a debt of the heart, or other personal obligations, many of them owed to family members. "Much of the passion, power, and loyalties ... are focused upon family," writes historian Alfred W. McCoy. The center of the society, family "defines life chances." He and others have used the phrase "an anarchy of families," which the anthropologist Robert Fox called the archipelago in 1959. Fox believed that the concentration of economic and political power within families weakened the government and political system, which in turn hindered economic development. Nearly a half century later, corporations, industrial sectors, political parties, and even whole regimes still depend on individual families and coalitions of families.[32]

As the logging industry boomed, it became, like the sugar industry, ever more linked with politics and the elite. Politicians, including senators, owned logging companies, served on their boards of directors, or had indirect interests in the industry. The government assessed negligible fees against the timber companies, whose taxes were estimated at less than 20 percent of what they should have been had the lumber's true value been assessed. In the 1970s and 1980s, despite economic costs that loomed if the timber industry failed, a number of logging bans, both provincial and national, were stonewalled in Congress. Log exports were not banned until 1986, more than a decade after the industry's peak.

The emphasis on personal relationships has continued to influence decisions and policies at the national level, and it also has distorted their effectiveness at the local level. The Philippines has scores of environmental laws that are decades old, many of them based on U.S. laws. They have remained largely unenforced, examples of what officials commonly call the "implementation gap" that separates regulations from reality. The branches of government that worked with natural resources were known as the most corrupt, and the forests proved particularly lucrative. There were endless possibilities for graft, starting in the field with the forest rangers and extending up to the highest levels of power. During the industry's heyday, the forestry agency, then called the Bureau of Forestry Development, was dubbed the Bureau of Forest Destruction.

The licenses that gave loggers access to the public forestlands also fostered logging. Timber license agreements were issued for only a few years, giving the timber companies little incentive to manage the forests for long-term yields. They tended to cut quickly and, other than building logging roads, did not invest in infrastructure. The annual amount they were allowed to cut was set high, and if they did not cut at least 60 percent of that amount, they could lose the concession. The coveted licenses also became key tools of political patronage, especially under Marcos, who dispensed the concessions himself, using them to reward family, friends, and supporters. In 1969, there were about 60 licenses; by 1976, there were more than 450, including short-term "special permits" given out as additional political favors. According to veteran journalist Marites Dañguilan Vitug, who received death threats for her reporting on logging, Marcos alone issued thousands of special permits. While a typical concession might cover 120,000 acres, those for his family and friends would be twice that size. During the height of the industry, nearly half the public timberlands in the country—about 12 million acres—were controlled by just over 100 individuals.[33]

Even the method of logging itself spurred cutting rather than conservation. The forest bureau stressed scientific management—selective cutting, described as the near-surgical removal of the largest trees. While this approach to logging might support sustainable forestry in temperate-zone forests, it has not done so in Philippine or other tropical forests. In theory, selective cutting should fell only the most valuable trees with the widest girth, leaving younger ones to keep the canopy intact and ensure the regrowth of the forest for future harvests. In practice, selective cutting removes far more.

According to a 1981 study by the Food and Agriculture Organization of the United Nations, logging operations in the Philippines on average stripped nearly one-third of an area of its vegetation.[34] Some of these trees were felled during the logging process; in dense tropical forests, where vines link trees' broad canopies, cutting a single tree can bring down its neighbors. Many trees were also cut during the construction of roads and tracks. One of the most destructive aspects of large-scale operations, roads can cover as much as 40 percent of a logging area. The heavy equipment used in large operations damages vegetation and also dislodges and compacts the loose soil, making it more difficult for new plant life to grow.

In some areas, loggers did not even pretend to cut selectively, clearing lowlands that were suitable for agriculture, just as Insular Lumber had cleared land in northern Negros for sugar plantations. Even where the lumbermen practiced selective cutting, they generally logged carelessly, breaking the limbs and damaging the trunks of nearby uncut trees. According to studies of logging practices, selective logging tends to harm far more trees than are targeted for cutting, leaving them prone to infection. One study showed that, while the trees left standing after a logging operation appeared healthy, at least half of them had rotted inside.[35]

◆　◆　◆

The Australian political scientist Peter Dauvergne has called the environmental effect that one country's economy has on another country's natural environment an "ecological shadow." In the Philippines, that shadow was cast primarily by the United States and Japan, which imported much of the forests that once blanketed the archipelago. Although the boom of the Philippine wood industry was initially driven by domestic demand and postwar reconstruction, it was sustained by the growing international demand. These exports typically were whole or raw logs, as they are called, the cheapest form of wood. Lumber, veneer, and plywood—and products such as cabinetry and furniture—all are more costly than the logs themselves. International demand supported the continued export of logs rather than the expansion of more profitable industries that could have ushered in greater economic development.

The United States was at first the biggest importer, but as Japan's industrial economy expanded in the 1950s, it imported more wood from the Philippines and elsewhere in the region, eventually becoming the world's largest buyer of

tropical timber. In the boom years of the 1960s, half of the recorded log production in the Philippines went to Japan, a trend that continued in the 1970s as
production slowed. Japanese general trading companies dominated the region's
log business, driving production and keeping prices low. They also worked with
smugglers and illegal loggers, which accelerated the most destructive cutting.
After exhausting the supply of accessible logs from the Philippines, Japan's loggers turned to other countries in Southeast Asia. Much of the deforestation in
the entire region has been attributed to the Japanese market. The tropical wood
that Japan bought, including the Philippine dipterocarps, typically was splintered into cheap plywood, used in construction to mold concrete, and eventually discarded.[36]

While buying Southeast Asian wood, the Americans and Japanese, having
long understood the long-term benefits of conservation and costs of deforestation, carefully managed their own forests. The Americans kept many of their
watersheds forested to prevent flooding and protect the water supply and, to a
large extent, safeguarded the vast national forests (although allowing logging
there, a practice over which debate has again heated up). The Japanese similarly
protected their own forests, fastidiously maintaining them as about two-thirds
of the country's land. During their otherwise brutal World War II occupation of
the Philippines, the Japanese brought some of their protective practices to the
country, limiting pastureland and prohibiting cutting trees near springs that
could be used for water supply or irrigation.[37]

By the late 1990s, Japan and the United States had lent the Philippine government hundreds of millions of dollars for programs designed to ease the ecological shadows that their demand for wood had helped cause. As in San Fernando,
Bukidnon, most of the replanting failed—because the funds fed the systemic
corruption and because the planting projects themselves were poorly conceived
and badly executed. The environmental loans increased the country's already
growing debt, which only added to the pressure to deplete the natural resources
to make payments on the debt.

◆

When Dean Worcester first encountered Mindanao in the 19th century, it was
the least known of all the islands he explored. The onion-skin map folded into
his 1898 volume labels the area of ocean just to the west of Palawan as unsur-
veyed and "dangerous ground" where ships were cautioned not to pass, but the
individual reefs drawn there still bear names, as do other islands throughout
the archipelago. Only on Mindanao do vast, unexplored stretches remain bare,
with no identifying names or markers.

In the 20th century, Mindanao's open frontier served as a kind of safety valve. The U.S. colonial administrators believed that expanding there could bring economic growth, as it had in the U.S. West; solve the problem of land shortages; and temper the unrest that the landlessness already was causing. Negotiating with the Vatican to buy up land formerly belonging to the wealthy Catholic friars, the insular government created a homesteading policy intended to placate landless farmers by offering them free passage to Mindanao. After independence in 1946, President Ramon Magsaysay used a similar approach to defuse the Hukbalahap rebellion in Luzon, which focused on the need for land reform. His homesteading policies lured tens of thousands of peasants to Mindanao, in places bringing in enough Christians to outnumber the *Moros* and stir up conflicts that continue today. As the interior of Mindanao was opened up, the widespread migration kindled the ecological destruction that over time would undermine the ability of longtime residents and migrants alike to survive on the land.

Long after the forests elsewhere had been logged, much of Mindanao's forests remained, making it possible for the logging industry to continue cutting without conserving. By the 1980s, Mindanao accounted for nearly three-fourths of the total area covered by timber licenses, two-thirds of all logs produced in the country, and nearly 90 percent of all plywood. These forests sustained the industry for the better part of the decade before they, too, had largely disappeared.[38]

Even as the forests dwindled, they remained an important source of income. Armed groups had long been involved with logging. Local elites often used private armies to guard their timber concessions, and various guerrilla forces were also involved. Marcos gave some of his special logging permits to Muslim rebels in Mindanao, and logging companies paid millions of dollars to the New People's Army as "taxes" to be able to continue to cut in areas it controlled. The police and military were also so often linked closely to logging, both legal and illegal, that their tendency to protect the loggers with M-16 armalites led the DENR to dub the practice "armalite logging." In Mindanao in the early 1990s, for a price rebel and government forces alike might safeguard the transit of timber, accept bribes to look the other way when logging trucks rolled by, or fell logs themselves.[39]

This source of income was not to be given up easily, as was the case in Guinoyoran, where Father Neri was killed. When Father Loloy Sangelan showed me the priest's small room, he and I stood silently, each waiting for the other. Danny's lessons had stuck. *Naninibago.* From Father Nerilito's room and the few possessions that he left behind, I understood something about the simplicity and focus of his life. From the town and other towns in Mindanao, I learned the unforgettable feel of a militarized place. For people living in these far-flung places, that feeling—from the CAFGU, the ever-present guns, the military detachments—could infuse their lives. Father Loloy and others I spoke with suggested that the sergeant suspected in the young priest's death was receiving a share of the logs passing through town. "Father Neri was killed because he was strong in his opposition to illegal logging," said Father Loloy. "That meant cutting off the military's means of livelihood. The military doesn't protect the loggers, but it has a share in the logs that are delivered. It's a sideline for them." The head of a major environmental group in Mindanao, who would agree to be interviewed only on the condition of anonymity, said: "The military was used to a lot of power under martial law. In many ways that power has not yet changed."

As he tried to fight the illegal logging in his parishes truckload by truckload, Father Neri might have thought about where the logs he confiscated would have ended up and what role international trade played in logging in Bukidnon. He certainly understood—in the intimate way of people who negotiated its dangers day by day—the corruption within the government. He also understood the lack of economic opportunity that drove local people, military or otherwise, to deplete their only source of wealth. In time, economic need became a central focus of programs to stem the illegal logging that persisted along the forest frontier, and that need turned out to be as intractable as the corruption within the government and industry and the vagaries and force of international trade themselves.

4 | Conservation with a Heart

Protecting land is an ancient tradition. From the storied forests in the Philippines to the castellated red rock formations of the U.S. West, indigenous people traditionally set aside land that they deemed sacred. As civilization encroached on wilder areas, the wealthy created private reserves where they could hunt. Refuges to conserve sites cherished for their natural beauty or unusual biology appeared more recently, burgeoning in number over the past decades as urban areas themselves burgeoned. Creating the protected areas posed only the first challenge; managing them effectively proved even harder.

When the Philippine government began revamping its environmental agenda in the late 1980s, there were hundreds of national forests, parks, and sanctuaries countrywide. So many of them existed only on paper that they were known as paper parks. People lived in the parks, and strict protection—essentially fencing off the land—could have only a limited effect. The communities might eventually benefit from conservation, but in the short term, unable to live off the land, they would barely survive. Most of the people had nowhere else to go, and if they were driven out, others would take their place. The only solution was to try to draw residents into conservation efforts and address both the communities' pressing economic needs and the need to protect the ecosystems around them.

The various protected areas were overseen by a hodgepodge of executive orders, laws, and presidential decrees, which were soon united by the National Integrated Protected Areas System. The system was created by a 1992 law, at the time considered innovative for both the Philippines and East Asia, that charged protected areas with the twin goals of sustainable development and conservation. One of the first pilot projects set up under the new law was the Conservation of Priority Protected Areas Project (CPPAP), which was funded by the Global Environment Facility, an international funding source affiliated with the World Bank. While such projects usually worked through the government, CPPAP forged an experimental partnership among the government, foreign funders, and nongovernmental organizations (NGOs). The seven-year, 20-million-dollar project was run jointly by the DENR and NGOs for Integrated Protected Areas, a national consortium of NGOs created to manage the project and dubbed NIPA, Inc.

CPPAP focused on 10 sites, selected from the last intact forests, marine areas, and wetlands, that were among the archipelago's most striking landscapes. These priority sites would be developed as models for conservation and local resource management. Rather than trying to protect them as wilderness, the project would establish elaborate sets of distinct zones, delineated by local people and scientists, to guide how the areas would be managed. The sites ranged from the scenic, landscaped islands of the northerly province of Batanes; to the Sierra Madre forest in Luzon, the country's largest refuge; to the breathtaking Agusan Marsh in Mindanao, where scores of species of birds winter. One of the most promising sites, also in Mindanao, was Mount Kitanglad, whose peaks are among the highest in the country.[1]

The Kitanglad Range covers northern Bukidnon Province, its steep volcanic mountains sweeping upward into the clouds. The word *kitanglad* means a place of *tanglad*, lemongrass, but the mountains are a place where the wild and cultivated compete. At their base stretch valleys planted closely with crops or, like much of the country, overgrown with *cogon* grass. Then the folded earth of the foothills begins, in places cleared so completely of vegetation that it looks scoured to bare rock. As farmers clear small plots, that bareness is creeping up the mountains, but the deep blue-green of the forest still spills up over the mountains' craggy silhouettes.

One of only two forests in Bukidnon that still have a closed canopy, the Kitanglad forest plays the vital role of maintaining the watershed for much of northern Mindanao. It also is distinctive as one of the most biologically diverse forests in the Philippines. Beneath the canopy live dozens of rare and endemic species, including the national bird, the Philippine eagle (*Pithecophaga jefferyi*), whose few remaining nesting sites include the Kitanglad range. The forest also shelters animals and plants once found throughout the archipelago: wild pigs and deer, monkeys, birds, bamboo, rattan, orchids, and ferns.

The rich forest life helped sustain the villages dotting the mountains. Populated mostly by indigenous people who retreated upland as migrants arrived from other provinces, these communities follow the road that loops around the mountains, beyond where the single line of phone wire ends. During the rainy season, the road softens into mud, turning the villages into temporary islands, although for many living there, the pennies that it costs to take the jeepney down the mountain make the trip unaffordable even in the dry season. The forest yielded food, building materials, medicinal herbs, and goods to sell in a place so separate from a cash economy that it has offered few other sources of income. Although the mountains were named a national park in 1990, established communities and new migrants alike continued to rely on their resources for survival.

A few years after the Kitanglad range became a CPPAP site, I visited several mountain villages to learn about the model for conservation being developed there. I stayed at a quiet, old hotel in Malaybalay, the provincial capital, and motored back and forth to the mountains with government foresters on their small Honda motorcycles. Together we crossed the gap between town and village life that posed one of the greatest challenges to the conservation program. In their bright offices in Quezon City, officials at NIPA, Inc., spoke highly of the park's strong superintendent and advanced program, which they viewed as one of CPPAP's most promising. Nothing they told me prepared me for the situation that I found on Mount Kitanglad or the real obstacles to conservation at the local level.

. . .

The Mount Kitanglad Range Natural Park is a huge area to govern. Covering about 77,000 acres, it falls into eight municipalities that fan out from a central ridge. Among them is Malaybalay, a small city with shops, the old hotel, and government offices. CPPAP was designed to accommodate a complex political landscape. For each site, it created a Protected Area Management Board, called a PAMB (pronounced pam-bee), a decision-making body composed of a protected area superintendent and representatives of all "stakeholders," from the government to indigenous and other local communities. The PAMB was intended as a democratic body that would include all groups in running the protected area, which otherwise might have been managed by the government alone.

Another cornerstone of CPPAP was the unusually prominent role played by NGOs. At the local level, NGOs offered the main link with the communities, which were supposed to be involved in developing and maintaining the parks. CPPAP also stressed the creation of economic activities that used natural resources in a sustainable manner. With the help of local NGOs, the communities would develop proposals for specific projects; nearly half of CPPAP's grant would be devoted to these projects, executed by the local NGOs and overseen by NIPA, Inc.

Kitanglad's 16-member PAMB included the eight town mayors, representatives from several NGOs and indigenous peoples' organizations, two regional government officials, and the park superintendent, Felix Mirasol. The recent devolution of power from the national government had given local governments more control over their own natural resources, but because the DENR managed the park, the municipalities could only help run it through the management board. Sometimes friction arose. The mayors were old-time politicians who often maintained power through force—"the goon type of mayor," said a high-level NGO official in Manila.

The mountain range's eight towns span diverse cultures, languages, and political and economic interests. The sheer size of the forest and the number of towns made managing the protected area challenging, as did the large proportion of indigenous peoples who were trying to claim parts of the range as their ancestral land. A community organizer commented, "One of the problems is that the people aren't all united—because of personal interests. There are conflicts in their faiths. They believe in different gods."

"The PAMB is said to be ambitious and unwieldy," said Felix, a forester in his early 30s with ample experience working with conservation efforts. "People think it can't be done, but it can be done, with patience and perseverance. The PAMB is powerful because it includes all the stakeholders. The only challenge is to get agreement. All the members are decision makers in their own communities, so it can be hard to get them together."

Villages and towns generally had their own distinct histories, which could still loom large. One town was abandoned when the New People's Army was strong and then repopulated when the military presence ebbed, but rumors persisted over who had fought with the guerrillas. Another town had experienced a rapid intrusion by migrants from other provinces. The towns, together with much of Bukidnon province, had endured a prolonged occupation by the Japanese during World War II, and people still remembered the hardships. Stationed on the forested and inaccessible mountains, for years some Japanese soldiers carried on the occupation alone, not knowing that the war had ended. In one community deep in the folds of Kitanglad, villagers still tell how soldiers eventually resorted to cannibalism, killing and eating a group of villagers, including a number of teenage girls. Even in the 1990s, this part of Bukidnon remained a world of its own, and rumors persisted that fresh sides of meat for sale in the markets might be human flesh. Although less isolated than in the past, the villages have remained places of innuendo and rumor where outsiders can elicit suspicion.

When I arrived, CPPAP was in its third year, its foundation for managing the mountainous park in place: the PAMB, Felix Mirasol and his staff, and a management plan. The map of the protected area included seven zones that ranged from the strictly protected core zone to the adjacent buffer zone, a transitional area where indigenous communities and long-standing migrants could gain tenure as residents. About 37,000 acres, or nearly half the size of the park itself, the buffer zone surrounded the canopied core zone. There was also a cultural zone where residents could pray and conduct rituals; a sustainable-use zone where people could use the natural resources, though less freely than before; a restoration zone where the forest, having been degraded by fire or overuse, was being allowed to grow back; and a recreation zone where people, mainly outsiders, could hike. While they could ascend most mountains freely, hikers had to buy permits and register at a checkpoint, and only 15 visitors could camp there each night. The fees would help fund maintenance of the park—a source of revenue that was still a novelty for local governments.

One of the trails leading up Mount Kitanglad starts in the village of Intavas in the town of Impasugong. I took several trips there with Edgar, a forester with the DENR. A compact man in his 30s who protected his arms from the wind's bite with a blue nylon jacket, Edgar was a thoughtful guide. He stopped frequently to show me a clear view of the mountains, hand-painted signs marking the park's boundary, the low level of rivers trickling through their channels. Uncomfortable with the trappings of working with a foreigner, he hesitated to enter the hotel lobby and declined even to eat in a restaurant, which made a day's travel something of an endurance test.

The road from Malaybalay to Impasugong was a wide span of dirt studded with stones that emerged during the rains. Edgar made the trip of 15 to 20 miles several times a week, and what usually took him about 40 minutes, he said, with me behind him on the seat took three times as long. By maneuvering his motorcycle along one of the two tracks worn by the wheels of the jeepneys and few cars that traveled the road, we could have a fairly flat ride, but it was a rough one.

At first we passed through pineapple fields, expanses of even, spiky rows stretching from the road almost to the foothills. The pineapples in Mindanao are small and unusually sweet. The plantations, run by international food corporations, offered some of the only jobs available, but they could be harsh places to work. Each pineapple plant is a clump of stiff leaves that arch into the next row, making the field a prickly obstacle course. Although the pineapple's jagged leaves can claw through cloth, workers received a single pair of overalls to protect their legs, and most could not afford more. Unless the thick fabric dried in the overnight heat, the day after they were washed, workers would have to sacrifice their own, thinner pants to the fields—and slice up their legs—or labor in damp clothes.

Beyond the pineapple fields, the road climbed into the foothills, narrowing into an irregular washboard of bumps. The motorcycle labored, its motor whining and its front wheel slipping from one fist-sized stone to the next. We moved so slowly that several times I lost nerve and hopped off. Edgar did not complain about having a passenger, nor did he ask me to stay on the motorcycle. He just looked ahead with great concentration. "This is my road," he said eventually. "I know it." He did know where to speed up and where to slow down, where to move toward the middle and where to steer to one side. Even when a brief, light rain wet the stones so that the wheels slipped more frequently, we did not fall.

After we left the fields behind, we started seeing people and houses and the signs for the Mount Kitanglad Range Natural Park. We crested a hill, and suddenly we were in the center of Intavas, Impasugong, an established village with fields, rows of concrete houses, and a store that sold bread, beer, and other goods. It was around 10 o'clock, and as we drove in, we passed young men squatting by the side of the road and alongside a small plaza. At a quick glance, they looked like men anywhere who are not working, men who wanted to be out of the house but had nowhere to go, and so they sat there with faces that seemed, between the laughter and talk that animated them briefly, blank with discouragement. Some of them were drinking. Some, said Edgar, were drunk already.

Because the village had no phones, we arrived unanticipated, but people were willing to interrupt their daily life to talk with us. With his easy manner, it was hard to tell whom Edgar knew and whom he was meeting for the first time. People spoke openly about the role that the forest had played in their lives, the creation of the protected area, and the changes that it had brought. We were escorted into one home nearby and then another. As I learned more, I began to understand, as I could not when I first saw them, why the men were squatting in the plaza and how they were connected with the park.

• • •

The trailhead was close, and from the village center, the road quickly became a steep path full of small boulders and ruts layered upon ruts where the feet of *carabao* had sunk into mud. Before long, the path narrowed to a rocky walkway that climbed into a forest thick with hanging mosses and lianas and great tree ferns whose fronds opened like fireworks exploding across the sky. The forest drew both local people and hikers, who trekked through what must to outsiders have appeared a quaint, peaceful village.

The creation of the protected area did not bar people from the forest, but it did impose restrictions on its resources, which had been a main source of income. There's no more logging, said Adrieno Bactor, a relative of the store owner. "There's no more collecting of butterflies and rattan, no more hunting of wild deer and pigs. People are not happy about this. If there were no ordinance, people would collect orchids and rattan, hunt the animals, cut the trees."

Enelson Sagayna similarly described a life formerly intertwined with the forest. A 38-year-old councillor in the *barangay* and father of seven children under 17 years old, he understood the value of saving the forest. "We protect the trees for the water now," he said. He was still able to gather rattan in the buffer zone to make furniture, but other activities, including farming, were curtailed. He had grown vegetables in plots cleared in the forest—which he called *kaingin* even though the farmers, mostly migrants, no longer followed the traditional practices, such as shifting cultivated sites, that once helped conserve the forest and its soil. Today, said Sagayna, *"Wala na"*—there is none of that. "There's no cutting of trees, no hunting, no gathering of butterflies and beetles. We earn about half of what we did before." And instead of harvesting the forest, the men worked temporarily as porters.

Portering did not mean, as I assumed, carrying hikers' packs and equipment. The president of a local porters' association, Sagayna described his work. On top of the mountain stood telecommunications towers owned by 14 companies and grandfathered into the protected area's management plan with a special-use zone. At the time of my visit, the National Power Corporation, Napocor, was rebuilding its facility. There was no road up which trucks could carry construction materials; instead, every day men loaded bags of sand onto their backs and carried them up the steep mountain path.

Some left at six o'clock in the morning, when it was still relatively cool, and returned about four o'clock in the afternoon. Others made two trips, departing at two o'clock in the morning, returning at nine o'clock, and then leaving again an hour later. For one trip, they were paid 6 pesos per kilo of sand, or about 180 pesos (then about 7 dollars) to carry about 60 pounds. Most of the porters were young and strong. Their bare, callused feet stepped quickly into the well-worn footholds among the tree roots. These men who served as beasts of burden were the men we saw along the road and plaza, taking a break in the middle of their 16-hour workday.

Although CPPAP emphasized providing alternative livelihood, in three years few such projects had materialized. Conflicts had arisen over the lack of work, the way that the zones had been created, and the restrictions on the forest. Many residents still found the restrictions in the protected area confusing, including the village *datu*, 56-year-old Carmelino Mahayao. A member of the

Higaunon tribe and president of both the tribal organization and the local consumer cooperative, the *datu* was a widower and lived with one of his six children. Edgar and I talked with him at their home, where we sat on the bamboo porch. The *datu* wore a dark nylon jacket, a Toronto Blue Jays cap, dusty blue high-top sneakers, and jeans. His hands, huge and strong with prominent veins, gestured restlessly as he spoke.

Even he did not fully understand the restrictions, blaming the protected area for a provincial log ban. "Before, the people would go and gather illegal materials in the park—wild animals, lumber. Today, because there's a national law, we can no longer go to the area and do what we did there before. Even for housing purposes, we can't lumber. That's the number one problem. Even in the buffer areas, we can't gather wood because of the moratorium on logging. If we had money, we could buy wood at the lumber yard—*coco* [coconut] lumber. Since the 1980s, some people also have planted *Gmelina* and *Falcata* [fast-growing species] and mahogany. They can get a permit to cut. But that's soft wood only."

As we spoke, the wind blew through the woven walls and kept us cool. "Before, I hunted wild animals—pig and deer—but only for my own consumption. I gathered wood, but only for my house. I gathered honey, which I can still do in the buffer zone. Some people still do these things, because it's our culture. They gather plants, which would be too expensive to buy. It's our culture. Different trees have different uses—to prevent a pregnancy, to end a pregnancy. Women gather them and mix with medicinal wine. We drink them to become strong or give them to a child after delivery. This is our medicine."

The Department of Agriculture had introduced a program encouraging farmers in the area to grow so-called high-value crops such as potatoes and cabbage, which brought more money than rice. In northern Luzon, farmers traditionally grew rice in irrigated, terraced fields ingeniously cut into the hillsides hundreds of years ago. In Baguio several hours north of Manila, these crops thrived in the cooler, high-altitude climate, and on some of the terraces straight rows of cabbage heads had replaced rice. The program to introduce them to Mount Kitanglad was controversial: In order to grow the vegetables, the farmers often cleared forest in the park. The vegetables, sold mainly to domestic fast-food firms, were lucrative, but they required more expensive inputs—fertilizers and pesticides—than traditional crops. In addition, the potatoes, typically grown on newly cleared land, would strip the land's nutrients. Already yields were dropping, in part because of depleted soil. When it came to their policies on high-value crops, several NGO officials told me, "the Department of Agriculture and DENR collide."

The *datu* grew high-value crops as a tenant farmer if he could get "financing"; if not, he grew the traditional crops of corn and *camote*, sweet potato.

With borrowed money, he could plant about an acre; without it, he would plant half that. "It's enough for me, but you can't send a grandson or granddaughter to school or build a house," he said. Like others, the *datu* cited the community's unmet expectations about how they would benefit from relinquishing access to the forest. They had believed that, by agreeing to forgo part of the harvest, they would gain other work to make up for the lost income. "We were convinced that the national park was important, because without it, the trees will be cut and maybe a flood will come. Without the park, no one can control the logging. I understood already the connection between the logging and the flooding. If there are no trees and there's a drought, then when it does rain, there will be a flood. I knew this before."

Since the protected area was set up, said the *datu*, their lives had not improved. He was disappointed at the DENR and management board. "Every time we conduct an assembly meeting, we have the DENR or PAMB brief us on the project, and they say that they will bring us livelihood. Until now, it's not yet implemented. There's nothing. Our house is only bamboo. The DENR and PAMB say that there's financing. So I'm waiting. If there's financing, we'll make a house, get a *carabao*. We're waiting. We're happy that we'll be given a livelihood. We'll be able to have a house. We'll have electricity." By this point, the *datu* was looking directly at Edgar, who was translating more slowly. "The problem is the government and NGOs go to our place with promises—like reforestation and livelihood. I'm still waiting, and there's nothing. Other agencies keep promising livelihood, but there's no effect."

As we spoke, a low cry kept coming from inside the home. When Edgar asked who was crying, the *datu* led us into a dark room. One of his grandsons lay on his back on a bed, surrounded by blankets and clothes. Sitting next to him, his mother barely looked up when we entered. The boy's legs were pressed together, one bent up, one stretched out. His arms were frozen at his sides. His eyes and face were clouded with misery, and he wailed steadily, like an infant, but much louder. "Bernito turned seven today," said the *datu*. "He had a polio vaccination a year ago, and his limbs froze. He can't see clearly or speak. Before, he used to run about."

We returned to the porch, and the *datu* spoke with rawness and intensity, moving his hands more. I could not distinguish his disappointment with the park from his pain about Bernito and anger at a government that did not support the poor. "The poor must be assisted," he said. "If people are sickly, they should be assisted." Edgar seemed increasingly reluctant to translate. "Every time they [the government] conduct a meeting, there are promises, and the promises are targeted at the poor, but there's no implementation. We have a health center here, but they only help with minor illnesses; with major illness, there's nothing they can do, and we can't afford it. If there's a *barangay* assembly and the *barangay* captain says we'll work without pay tomorrow for the good of the *barangay*—to construct a basketball court or something—we'll do it. Whatever he says, we do. But the needs of the people, like food, can't be given to us. And in terms of sickness, if we ask assistance, we get nothing. There's no time to address our problems in the meetings. They always adjourn at two o'clock, and we've had no lunch, so there's no time to ask questions. We help the *barangay*, but the *barangay* won't help us."

⚡

The first modern protected area was created in 1864 to safeguard the craggy landscape and great sequoias of Yosemite Valley in the United States. The first U.S. national park followed six years later to protect the forests and thousands of geysers, hot springs, and other geothermal wonders at Yellowstone. Although protected areas like these originally had a narrow goal of safeguarding striking landscapes for recreation, in time they became a global model for protected areas in general. Conservation came to be equated with the preservation of wilderness, lands envisioned as untouched, uninhabited places that needed to be shielded from practical use and buffered from the presence of people. Yosemite itself was guarded by the military until the creation of the National Park Service in 1916.[2]

Protected areas are often based on a classification system developed, as part of a campaign to promote and track global conservation efforts, by the Switzerland-based International Union for the Conservation of Nature and Natural Resources (IUCN). Also known as the World Conservation Union, the IUCN is considered the international arbiter for matters of conservation. Founded in 1948, today it is composed of more than 800 government agencies and NGOs from more than 125 countries. Its categories of protected areas are based on a "gradation of human intervention." Some areas are set up for strict protection, while others are managed for a combination of protection, recreation, and harvesting of their resources. They range from wilderness areas to national parks to reserves whose resources are managed for long-term use. One IUCN category describes a resource reserve intended to curb development and conserve the natural resources for future use, although few countries have created this type of conservation area.[3]

While by definition people are considered outsiders in the IUCN protected areas, there actually are few places in the world that humans have not in some way altered.[4] Yellowstone itself—whose apparently pristine beauty inspired the creation of reserves worldwide—has been inhabited for at least 11,000 years.

At the time it was designated a national park, the Shoshone and other Native American tribes hunted and fished in the highlands during the summer, and their use of fire helped shape the memorable ecosystems.

Although early administrators at Yellowstone may have pressed for the tribes' exclusion for security reasons after an assault on tourists in 1877, keeping the area uninhabited was not integral to the early vision for the park. The area clearly had great potential for tourism, and the U.S. Congress created Yellowstone to prevent its privatization and destruction. The word *wilderness* was not mentioned in the congressional debates or legislation. The tribes stopped living in the park year-round by the late 1870s, but it was most likely their confinement to reservations, rather than exclusionary policies, that drew them away from Yellowstone.[5] Many protected areas, in fact, have been created in indigenous people's homelands, precisely because they tend to be relatively undeveloped areas.

In the 1960s and 1970s, the rationale for protected areas began to change. With mushrooming populations and cities, the widespread "conversion" of land meant that whole ecosystems were disappearing. Scientists began to speak publicly about the loss of these natural areas and a sudden increase in extinctions. Although extinctions are clearly part of evolutionary history, the scientists believed that the current extinctions were greater in number than before, that they were unprecedented, that they were irreparable, and that they would prove to be catastrophic. Environmental historian Donald Worster, in tracing 150 years of ecological thought, captures their concern: "Almost anywhere one looked in the plant and animal kingdoms, the picture was getting very grim: an acceleration of extinction so great that it amounted to a reversal of the processes of biological evolution."[6]

An adamant spokesman for the scientific community was the biologist Edward O. Wilson, who coined the word *biophilia* to describe the "innate tendency" that humans have to "focus on life and lifelike processes." Wilson questioned what the extinctions might mean. "What event likely to happen during the next few years will our descendants most regret?" While he acknowledged that the worst scenario would be global nuclear war, other catastrophes could be "repaired within a few generations. The one process now going on that will take millions of years to correct is the loss of genetic and species diversity by the destruction of natural habitats. This is the folly our descendants are least likely to forgive us."[7]

In industrialized countries, the environmentalism of the 1960s and 1970s succeeded in reducing industrial and urban waste as well as protecting wild and unique ecosystems. Rural areas were disappearing under asphalt and subdivisions, and the public responded to scientists' fears about extinction. Even if they could not stop the proliferation of suburbs, shopping malls, or vacation homes, they might be able to spare natural places where the forward momentum of development was less relentless. In Southeast Asia and Latin America, tropical forests were going up in smoke as land was logged or cleared for crops. These forests became a focus for western environmental efforts. Although years earlier they might have seemed the pinnacle of discomfort—hot, humid, and full of parasites and winged insects—tropical forests were embraced for their incredible diversity of life. The number of species in these jungled forests remains unknown, but they are believed to harbor at least half of all species alive today.

Along with the concern about extinctions came a new term, *biodiversity*. The word was collapsed from the phrase *biological diversity*, which when I studied botany was a technical term for the average number of species in an area. Biodiversity was a word that the public could understand, a way to separate biological life from humans' use of it. It had a broad definition, covering biological systems at all scales: genes, species, ecosystems. Early on, exasperated skeptics questioned how, without an exact definition of biodiversity, we could promote its protection. For a public beginning to understand that environmental problems are global and that environment and development are intertwined, however, biodiversity was more concrete than declining ecosystems or genetic resources. "Ecologists began to argue," wrote Worster, that "whatever the uncertainties of theory, we must prevent the extinction of any and all species of plants or animals at the hand of man."[8]

Concern about declining biodiversity—in tropical forests, coral reefs, and less exotic and diverse ecosystems—influenced conservation. The Convention on Biological Diversity, a key agreement adopted at the 1992 Earth Summit, committed countries to conserving their biological resources and using them sustainably. According to the IUCN, in 1962 there were fewer than 10,000 protected areas worldwide. By 2003, there were more than 100,000, about half of them established in the previous 10 years. The original protected areas created in wealthy countries were primarily dedicated to recreation and the preservation of national monuments. The newer areas, established mainly in Africa, Latin America, and Asia, protected sites with extreme natural beauty or diverse and endemic species.[9]

As the numbers of protected areas grew, ideas about how to make them effective also evolved. In the late 1980s, the U.K.-based biologist Norman Myers, who raised an early warning about the destruction of tropical forests, proposed that the first priorities for conservation should be areas that had the highest rates of endemism and were declining quickly. Myers called these threatened areas genetic "hot spots." They deserved attention, he argued, because while they made up less than 3.5 percent of remaining primary forests, they contained more than 34,000 plant species. That meant that these hot spots harbored 27 percent of all plant species found in tropical forests and 13 percent of all plant species worldwide. They also contained at least 700,000 endemic animal species, mostly insects. He named 10 sites that needed immediate intervention, including the Philippines.[10]

Another approach has been to focus on areas with high rates of biodiversity that are so remote that they actually have a good chance of surviving relatively intact.[11] Larry Heaney, a biologist at the Field Museum of Natural History in Chicago, questions whether we need to choose between the two. Since the early 1980s, Heaney has done research in the Philippines, where he identified 18 new species of mammals. While many tropical biologists avoided practical problems related to disappearing ecosystems, Heaney got involved, training Filipino biologists and working to reestablish protected areas, including Mount Kitanglad. Rather than choosing between areas most and least threatened, said Heaney, "We need to do both. I don't care how bad things are; philosophically and ethically, it is wrong to give up." He says that it is also wise to conserve a range of areas. In the Philippines, he added, species are fairly evenly dispersed among the islands and require different protected areas. "You can't have just one park in Luzon and one in Mindanao—you won't get the species."[12]

The increase in protected areas paralleled a growing resistance to large-scale projects, such as dam construction, and a global movement to preserve indigenous peoples. Setting up parks that excluded an area's inhabitants, as was done in the United States, was no longer acceptable—or feasible. Remarked the international conservationist David Western, "If we can't save nature outside protected areas, not much will survive inside."[13] Conservation became linked with communities and with economic development that would allow the long-term use of natural resources—the now familiar notion of sustainable development. Protected areas would be managed not only for strict protection, but also for a range of purposes, with the hope that involving communities in conservation would make them more receptive to its restrictions.[14]

As on Mount Kitanglad, the traditional systems once used to manage communities' resources had been largely abandoned or forgotten. Where the wealthy have had disproportionate control over the natural wealth, as was often true for forests and fisheries, people tended to compete to exploit it. Managing resources for long-term use requires social and political development—coordinating various levels of government with NGOs representing a range of interests and needs. It requires new policies and laws, long-term funding, the participation of scientists, and a skilled workforce. It means creating new ways for governments and communities to work together. It also means teaching generally poor people about the long-term benefits of conserving their base of natural resources. In the Philippines, CPPAP pioneered this approach.

⚡

A second trailhead into the forest's canopy begins in San Vicente, Baungon. San Vicente also lies at the end of the road, but unlike Impasugong, it is just a bit of a village. Set on a hill, it overlooks a broad valley where, even within the protected area, bare patches have been cut into the canopy. The forest begins a few hundred yards from the village center, where a steep path leads up the mountain. Tourists seeking that trailhead were a common topic of conversation among villagers, especially when they spoke with outsiders. Many foreigners come here; they find our place beautiful, one resident offered. To distinguish the hikers from his own, small-boned self, he cut the air with his hands, outlining their height and the packs lashed, like upright turtles' shells, to their broad backs.

It does not take long in Baungon to see how hard it can be to convert a forest into a protected area. Many residents regarded the park as a place for outsiders or a place to protect the Philippine eagle; they felt excluded from its creation, and conflicts within the community itself limited the conservation efforts. Despite good intentions, despite adequate funding, despite the involvement of local NGOs and community organizers, the residents largely continued to use the forest as they always had.

In the late 1990s, most people living in San Vicente were *lumad*. "This is the cultural zone," said Marvic, a determined community organizer from the Haribon Foundation who had been working and living there half-time for three years. About one-third of the residents were migrants, who were still streaming into Baungon. The hope for CPPAP had been that if residents were given secure tenure, they would be more likely to conserve the forest, and their presence would deter others from moving there. Instead, the migration continued. "The *barangay* captain invites migrants," as a way to garner votes, said an NGO worker.

A tiny, 50-year-old woman stood near her home, a blue tarp hung over a waist-high line not far from the main path. Hernina had four daughters, all married and living several hours away on the coast in Cagayan de Oro, the largest nearby city. Her husband had come to Baungon first; she had arrived three weeks before. "I'm a *lumad*, but I'm new here. I lived in Cagayan before. I came because of relatives here. My husband farms: *camote, gabi, camoting gaboy, mais, marang* [sweet potato, greens, cassava, corn, fruit trees]. We've just started, so it's hard—*mahirap*—very laborious. But it will sustain our income."

Her husband's farming plot likely was one of the bare patches that shone through the forest's canopy. As in Impasugong, there were few ways to feed a family here. Decades earlier, three logging concessions had operated in Baungon, providing jobs for residents, who also did small-scale cutting. When the military conflicts escalated, many left; when they returned, logging had been banned in the province. The small cutting continued illegally, but creation of the protected area soon followed, which further restricted logging and other activities.

Some of the tensions over Mount Kitanglad had to do with the cultural gap between the *lumad* and lowlanders. "People feel inferior," said *Datu* Vic Saway, a leader of the Talaandig tribe who lived in Lantapan, where he had grown up. When he spoke of the indigenous people, he used the common shorthand, IPs. "Most IPs think that [the lowlanders] are better and won't stand up to them," he said bluntly. A member of the protected area's management board, *Datu* Vic believed that there was a strong relationship between the survival of groups such as the Talaandig and conservation. There were three divergent concerns on Mount Kitanglad, he said: the DENR was concerned with environmental protection and conservation; the local governments were concerned with development; and the IPs were concerned with cultural survival.

"There's a relationship between the environment and cultural survival that some people don't understand. The cultural factor is important to the environment, and it's important to development." He did not pretend that the indigenous communities had retained their traditional conservation practices, but he did believe that they were sufficiently intact culturally that, if given control over their ancestral land, they could manage it well, and in a less polarizing manner. "Development programs shouldn't divide a community, as they do now," he said.

Other tensions had to do with the protected area's zones. While CPPAP called for the communities to participate in mapping out the zones, reports differed on how involved they had been. According to protected area superintendent Felix Mirasol, the DENR had consulted the communities. "We held public hearings in the *barangays* and *sitios*. We held workshops with the local governments, the hunters, the vegetable growers, even the smugglers, because they'll be the future problem. Different people from the same municipality were given maps and asked how they would assign the zones, asked about the soil, slope, vegetable cover. The next day, people showed their map to the group and explained it. Some of the main problems came from the political boundaries; we couldn't join them." Without accurate maps, people disagreed over the towns' boundaries. "A committee with three people from each of the eight municipalities met and made a map," he said. "It was presented at hearings, and we heard the concerns and consolidated them."

Haribon's Marvic was concerned that the *barangays* had not been involved equally. "There was supposed to be a public hearing, but the DENR never came here." She pointed down the road. "Maybe they went down there"—to villages lower down the mountain or Baungon proper—"but they didn't consult the people who really make a living from the forest. When you talk about a protected area or a national park, the *lumad* doesn't like it. They were never part of the plan, and they don't know about it. They don't know what a protected area is; all they know is that when a place is a protected area, they can't go there."

The biologist Danny Balete, who has collaborated with Larry Heaney for many years, worked on mapping the zones on Kitanglad. "We recommended zones, based on species and habitat conditions and the way that people used them. Zoning is for use. But the NGOs didn't really have a good survey of the use of the area. You can't just show people a map and ask them how far into the forest they're collecting wild pigs and such. They held a one-day meeting and called it community mapping, but is that what it was?

"Establishing management zones is really a negotiation; you have to bargain with the people: given the benefits of this zone, how big should the strict protection zone be? You can only do this if people understand the zones and their consequences. Even though some of the NGOs worked with the communities at first to map out the management zones and explain their implications, people didn't really understand the zones; they didn't even understand the terms."

Having worked to revive the Philippines' system of protected areas, Balete understood that mapping should be closely linked with organizing, which requires a long-term commitment to a community. It also involves learning what the community needs, not just preparing it for what outsiders might want. "You have to go in and ask people what they want, such as a road or irrigation," Balete said. "If they're organized, you can help them work with the different government agencies and get the service that they need. You even start before that, start with the interpersonal conflicts and give the community skills, help them function as a community." Organizing means education. "You need to invest in the people. They may be the best managers of their own resources, but in order to do it, they need the skills. You don't learn how to facilitate meetings by planting rice or trees."[15]

Although a cornerstone of CPPAP, community organizing turned out to be difficult, despite ample experience elsewhere with environmental projects. Rather than working with the communities early on, Balete said, "[Organizers] would just go to the areas and say that they had to organize because there was a fund." Ironically, part of the problem was the ample funding. The main incentive used to persuade the community to stop harvesting the forest was financial, and in such poor communities, the money, rather than conservation, quickly became the focal point. Heaney, who created a "classroom in the field" on Mount Kitanglad to teach scientists and NGO and government workers about biodiversity and conservations, agreed: "Too much money is the worst. It promotes corruption that feeds the source of the problem. CPPAP had too much money, and a lot of it was mismanaged or it stayed in the provincial capital."

The organizing took longer than expected, said one of the top officials at NIPA, Inc., which itself was problematic. Balete explained, "Funding agencies don't necessarily want to fund community organizing, because you can't set a timeline or specific goals for that. They want to see specific deliverables: how much money was spent, and for what. Organizing may show little results for years; the funders don't realize that it's the most important part."

Organizers from the DENR and NGOs found working in Mount Kitanglad's remote communities to be particularly challenging. Unlike organizers I met elsewhere, those in San Vicente described experiences that conflicted sharply with their training, their other experiences, their expectations of what *should* happen. "We conduct dialogues and trainings, teach them the importance of the environment and about alternative livelihoods. We do hands-on demos," explained a DENR organizer. "But some people think the meetings are useless. They don't know how to write. Also, they have to work in other places to buy food." As we stood talking along the road, a man passed carrying a load of rattan from the forest. No one questioned him, but when asked why people still gathered wood, a woman new to the area hesitated, then said, "When you farm, it's really hard. It takes time before you can earn money. If you cut trees, it's easy to get money."

A father of three who lived near the road had weathered the economic transitions in Baungon that had made it progressively harder to support his family. Although he was only 27, he looked far older "because of the responsibilities," he said. "Before, I did illegal logging because of financial need. There were no fields then. Because of the presence of the government here, I stopped, but we don't have a *carabao* to plow or money to buy fertilizer."

Like other residents, he viewed the DENR organizers as government representatives, which limited their effectiveness. "We wish the government would allow us to use some of the forest products—not for the main source of our income, but as an alternative livelihood." When asked whether people were angry that they could not use the forests, he replied, "Not angry but *malungkot* [sad]." As if not wanting to offend, he added politely, "If the government would give us livelihood, it's better."

The DENR organizer agreed, citing the well-known aphorism about social change: "It's better to teach to fish, not to give fish." But he acknowledged that teaching people how to fish—giving them a way to earn a living—was not straightforward. "The people don't understand the importance of the protected area—they think it's still okay to use the forest."

He understood that the *lumad* culture had splintered and that the presence of outsiders had changed the traditional ways. "Before the loggers came, people hunted and fished, but then the concessionaires taught them that the trees could earn money for them. The people are also influenced by the migrants—who buy the forest products from the *lumads*, buy land from them." Sometimes they sell the land over and over, he said. In coastal areas where he had worked previously, the fishers were more educated and had been easier to organize. On Mount Kitanglad, people understood what the organizing taught, but it did not seem to make a difference. "They only tolerate, they don't follow," he said. "Some are lazy. They prefer to do the illegal activities because it's easier."

"It's difficult," he conceded. "The road isn't passable, it's expensive to travel, people don't know how to garden [how to farm], and farming is expensive. In order to organize them, you have to do immersion—stay with the people. You have to teach them."

While such immersion might have worked in other communities, for outsiders Baungon was a place short on trust. Marvic, the Haribon organizer, remembered, "People were suspicious of us when we first came. They would close their doors when we approached." Even after time had passed, the tensions continued. Organizers from both the DENR and NGOs could talk with confidence and conviction about the challenges of organizing and then suddenly and emotionally blurt out the intractable problems that they faced.

"They insult us sometimes," said the DENR organizer. "There's backbiting." In fact, several villagers complained to me privately that he himself only lived in Baungon a few days a week and often found reasons to stay away longer.

The organizers did not even trust each other. Although they arrived around the same time, they had set up their projects separately. "They didn't coordinate with each other," said a NIPA, Inc., official. "They implemented separate but similar projects. There was a lot of suspicion on both sides." The competition among the groups trying to organize within Baungon created a palpable tension, and several of the organizers seemed overwrought.

I spent the night in Marvic's small wooden house and saw San Vicente grow even smaller. After the sun set behind the mountains, blackening the valley, children played alone in the road. Illuminated by the Milky Way, they ran around lighting bits of paper and whirling them to make spirals of gold that shone briefly before vanishing. Men walked nearby in pairs, their gait slowed by drink. Inside, Marvic lit a lantern whose glow passed through the shutters as stripes of light that lengthened on the ground as she carried the lantern across the room. I soon joined her and three young men talking in her living room.

Although the men were shy with me—"I've run out of English," said one—they spoke frankly about the "jealousies" in their community and about their role as *lumad*. A project planting wild seedlings in a partially logged area had received generous funding from a foreign foundation. "The beneficiaries were supposed to be *lumads* from this *sitio*," said another, "but others have benefited as well. If I push a *lumad* idea, I'm not heard. No one listens to our suggestions." The DENR had told them that the officers needed a "technical background. But most of us haven't been to school. The problem is that all the officers are in one family," the one family whom the funding mainly benefited. While the project involved planting trees, it also involved managing the area; the *lumad* did not have much experience, but he believed that they could gain the necessary skills. "We need leadership seminars, management seminars. Even if we're illiterate, we can learn."

Marvic advocated for the *lumad* and helped them negotiate the complicated relationship with the government, but her project was funded only through the end of the year. "What will happen to these people when I leave?" she asked. She had been working there long enough to earn the trust of many people, but she also had absorbed the tensions of San Vicente. Before we blew out the lanterns, she seemed afraid, with the fear that I saw all along the forest frontier.

◆ ◆ ◆

Not long before CPPAP ended, I spent several days with Danny Balete and Larry Heaney at the Field Museum of Natural History in Chicago, where they were completing a project. A slim man whose glasses do not mask the warmth of his eyes, Balete has devoted much of his life to conservation. Sitting at his lab bench, we were half a world away from Mount Kitanglad, but his intensity brought the mountain villages near. Before long, in our minds we were back under Kitanglad's canopy, seeing the arcing fronds; the small, quirky mammals that Heaney's team had discovered; the porters stepping down the rocky path, averting their eyes from outsiders out of shame or shyness or both.

Balete grew up on a farm in Bicol, a very poor area in southern Luzon. A good student, he received scholarships to college. "I was ambitious. There I was, the son of a farmer, and I wanted to be a scientist. I wasn't aware of conservation problems." After he finished school, his dreams changed. Haribon hired him to work on the initial mapping of protected areas in the 1980s—the first step toward revamping them. "That's how I got started. I soon realized that the

rattan. It's true that if you solve the problem of the people, you've solved most of the problem, but the thinking was too simplistic." When he spoke of "people," he meant those living in the communities; others he called "outsiders."

Having grown up in Bicol, Balete knew about poverty firsthand. His mother still lived there with three unmarried sons, who had only gone to elementary school. His brothers "sit around the house and don't work unless they have to," he said. He would visit, but he could only do so much. "Our lives are so different now, but it helps me to understand people's perspectives. I understand the poverty and the difficulty of their lives, the way that they have so few options; I understand their immediate need to feed and support their families."

Balete stopped and took a breath, as if startled at having spoken at such length. He still believed in the feasibility and importance of the protected areas, but to work they had to be humane. To succeed, resource management—and conservation—had to accommodate the reality of people's lives.

Accommodating rural life partly meant not romanticizing it. "People want better lives. A lot of outsiders coming into the communities don't realize this," he said. They do not realize that the poor are quite aware of their poverty. "People say this to me: 'Who wants to walk for three hours to gather a bundle of rattan that they can sell for a little money? Who would want a life like that?' They're aware; they know. They so much want the opportunity to improve their lives, but we need to work with them to help make that happen.

"People equate conservation with enforcement, think that it only prohibits them from doing anything. A migrant in the protected area in the Sierra Madre asked me, 'Is there any conservation with a heart?' He thought it was just the DENR going around with a big stick and whacking them every time they crossed an imaginary line into the protected area."

Part 11

The Coasts

5 | Watchdogs of the Sea: *Bantay Dagat*

Sometimes a single stretch of horizon captures something basic about a place. In the rural provinces, that landscape is a seascape, the silhouetted edge of an island broken by pinpoints of light just where the land slips into the sea. In the late evening, the water lapping toward the shore is flat and luminous. A few *nipa* houses face the shore, overlooked by sturdy coconut palms and papayas. Nearby, bright wooden *bancas*, fishing boats, have been drawn onto the sand, lined up like animals asleep for the night. By dawn, the boats have headed toward the coral reef, which glows like an aqua ring around the island. Their bamboo outriggers slapping the water, the *bancas* ride low as they pull forward, carrying one or two men who tend the motors or paddle evenly. On larger boats, awnings shelter the fishermen, but on the smaller ones, the men work under the full sun.

The forests may be the archipelago's ecological anchor, but the coasts provide work, food, and a way of life. It would be hard to overstate their importance. With more than 7,100 islands, the archipelago has one of the longest coastlines of any country in the world—about 11,000 miles, approximately the distance between Manila and Buenos Aires. About half of the population lives in the thousands of *barangays* lining the coast. After rice, seafood—including fish—is the most important part of the diet, providing about two-thirds of all animal protein, particularly for the poor. In 1996, the country was the twelfth-largest fish producer in the world, exporting much of its shrimp, prawns, tuna, and other seafood to Japan and the United States. The official annual harvest of 2.3 million metric tons—as with logging figures, a rough estimate at best—is divided among small-scale fishing, aquaculture, and the largest sector, commercial fishing. More than 1 million people work in the industry, about two-thirds of them as subsistence or small-scale fishermen who fish the near-shore reefs, ecologically the richest fishing grounds.

Despite their importance, the fisheries are contested areas that are hard to protect and manage. Like the forests, they are being degraded by large and small users, and they face pressure from a huge influx of migrants. The coasts are also the site of conflicts over what kinds of development should be pursued—from heavy industry to ecotourism. Marine ecosystems are complex; while resilient in many ways, their fisheries can collapse, as is happening worldwide. Scientists cannot predict when and why fisheries fail, and they also do not know what makes marine populations revive. While the forests tell the history of the overuse of the country's natural resources, the coasts offer important examples of how complicated decisions about conservation and development can be made and what makes conservation succeed over time.

In the Philippines, nearly all coastal fisheries have been overfished, and the ecosystems that support them are steadily eroding. A survey by the University of the Philippines Marine Science Institute showed that by the mid-1990s, three-fourths of the coral reefs were in fair to poor shape. A square kilometer of well-preserved reef yields on average 15 tons of fish a year but can yield up to 37 tons; a damaged one may yield one-tenth as much. The pressure on many fisheries has surged to at least twice what can be sustained, and stocks of bottom-dwelling fish are about a third what they were in the 1940s. As their catches decline, fishermen work longer hours and fish farther from shore. With more people fishing—and fishing harder—overall production has continued to grow, with commercial boats taking an increasing share. Nonetheless, according to the Asian Development Bank, from 1983 to 1994, catches—fish landings—from coastal waters in the Philippines dropped more than 11 percent.[1]

The once-rich fisheries are declining for the same reasons that fisheries around the world are in crisis. First, improvements in gear and increases in the numbers of both small and commercial boats have heightened the pressure on fisheries. As catches dropped, fishermen fished harder and sought out other fisheries. Second, fishermen also increasingly used destructive fishing techniques, particularly dynamite and poison, which boosted their catches but destroyed the fisheries and reefs. Third, the reefs and coastal fisheries have been damaged by sewage, pollution, and runoff from logged areas. Finally, coastal management has repeatedly been stymied by neglect and by denial that the ocean's bounty could ever be depleted.

Like the forestlands, fisheries used to be controlled at the national level, but because the country has such a long coastline, there was minimal enforcement of fishing laws. As with logging, wealthy families and politicians also played

Street boys on a cruise. Mindoro.

important roles in the industry, further stymieing enforcement. Several recent laws shifted management of the coasts to the local level. The devolution of the national government in 1991 gave each local government responsibility for its own municipal fishery—defined as the 15 kilometers of water adjacent to the shore. The 1998 fisheries code reserved these fisheries for the small-scale fishermen, who use boats of three tons or smaller, often *bancas* with simple gear, or pumpboats with a motor sunk in the center. These laws make it easier to restrict gear, catches, and access to fishing grounds. In an effort to revive their fisheries, control fishing, and curb illegal fishing, hundreds of coastal villages have set up marine protected areas, small no-take zones created and enforced by the community that protect a stretch of coral reef.[2]

One of the oldest and most successful reserves began as an experiment in the late 1970s on Apo Island in the Visayas. Although Apo's marine sanctuary has faced new threats from developers and tourists drawn to the reserve's flourishing corals, it remains a model for other villages and towns, its name almost synonymous with coastal management. Apo also draws visitors from all over the world interested in how small reserves can enhance neighboring fisheries. This approach to conservation, pioneered in the Philippines, offers hope that the emptying of the oceans might be reversed.

In places such as Apo, fishing is the occupation of last resort. While overfishing shortens the economic life of fisheries, it allows the fishermen and their families to survive. They live so close to the margin that any conservation effort can appear threatening—a hard fact of coastal life. A biologist working with fishing communities on the northern coast of Negros Oriental explained, "At first, the fishermen didn't even like people coming and talking with them about conservation. For them that word is taboo. When we first came, we asked a number of questions—including, Are you in favor of conservation? Are you in favor of setting up a reserve? They did not like these questions; they worried that they would be deprived of their livelihood. We learned that you need to start from the bottom—start without mentioning a reserve at all."

• • •

A volcanic island of about 180 acres, Apo Island lies at the southern tip of Negros Oriental, about 16 miles from Dumaguete City. With a thoughtful oceanography student whom I will call Pabs, I took a surprisingly bumpy road south from Dumaguete to Malatapay, Zamboangita. Malatapay seemed a sleepy village, but fishermen from the more scenic Apo rely on its weekly market to sell their catches. A pumpboat ferried us to the island in about 30 minutes.

Approached by boat, Apo has an uneven profile, with a high hill at the northern end, a low hill at the southern, and cliffs that plunge to the sea. Five short white sand beaches and a coral reef close to shore draw divers and other tourists, but despite its beauty, Apo can be inhospitable. During the dry season, the sea intrudes into the groundwater, and residents must collect rain or haul water from the mainland. The hills are stony and steep, and no more than a third of the island has flat, fertile soil that can be farmed easily. Most of the trees were cut long ago, and some of the hillsides are overgrown with *cogon*, which, once established, is hard to oust, as on the northern end of Apo, called Cogon Point.

The coral reef surrounding Apo supports what historically has been a rich fishery, particularly off Cogon Point. The fish are diverse and abundant, in such economically important groups as snappers, emperor breams, sturgeons, jacks, groupers, rabbit fish, and parrot fish. In the 1990s, about 600 people lived on the island, with 200 fishermen, some of them migrants from other provinces. The fishers used simple gear: hook and line, nets, and spears. The water was so clear that they could reportedly free-dive up to 20 meters to spear fish. Although Cogon Point was fished hard, yields had remained high—partly because of the reserve, and partly because it could not be fished year-round. During *habagat*, the southwest monsoon from May through September, the sea is calm. However, during *amihan*, the northeast monsoon from October to March, high winds keep boats grounded.

Life in coastal fishing villages is spare. According to surveys conducted for the Asian Development Bank in 1989 and 1990, about half of all fishing communities in the archipelago lacked electricity in homes, only one-fourth had running water, and only 4 percent of the fishermen owned beds. On average, fishers had barely five years of formal schooling, and about 80 percent of the families lived below the poverty line.[3]

Apo Island is typical of fishing villages. Even in the mid-1990s, the island offered no schooling past the elementary grades, and except for one midwife, there was no medicine or health care available. About three-fourths of Apo's families reported having lost at least one child, many in their first year. Although the reserve had helped restore the fishing catches, to make ends meet, in one-fourth of the families, at least one member worked off the island, sending money back—especially during the lean *amihan* months.[4]

⌐

Coral reefs are the rain forests of the sea, unparalleled in their spectacular display of life. Together with Borneo and New Guinea, the Philippines forms a side of what is called the Coral Triangle, the area where the Indian and Pacific Oceans flow together. The Coral Triangle contains more species of corals, fish, sea grasses, and other marine life than anywhere else in the world, a diversity that decreases in a steady gradient away from the equator. The archipelago has about 27,000 square kilometers of coral reefs, which range in length from a few yards to three miles. Most of them are fringing reefs, which absorb the force of waves and typhoons. Scientists have identified about 400 species of hard corals and 1,000 species of fish in these reefs.

As with the forests, warnings about the need to manage the fishing grounds went unheeded. Natural resource policies treated the ocean life as infinite, even though published reports of declining catches date back a half century. A 1957 issue of the *Philippine Journal of Fisheries* describes overfishing in the waters of the Central Visayas, which had accounted for more than one-third of the country's harvest. According to the author, who worked for the Bureau of Fisheries, in the early 1950s trawlers—large boats that drag fishing nets across the ocean floor—could get a full catch in a single day in the Guimaras Strait between the islands of Panay and Negros. Within a few years, trawlers had to spend five days and travel as far as Masbate. "Casual examination of their catches," reports the journal, "also showed the predominance of immature fishes," revealing that even then bottom-dwelling fish were "undergoing depletion."[5]

In the early 1960s, the economy was growing, and the country's natural resources still promised wealth. The government promoted the fisheries to

other countries as a new economic frontier—"untapped," "vast," and "rich"—
and the industry continued to expand.[6] International prices for fish and shrimp
were rising, and as a form of coastal development, a government program oversaw the replacement of mangroves bordering islands with aquaculture ponds.
Into the early 1970s, national policies continued to support the accelerated
exploitation of the oceans, despite growing evidence of overfishing and declining catches. As late as the 1980s, to boost production, foreign aid was used to
motorize fishing boats, upgrade gear, and expand fleets.

As catches declined in the mid-1970s, commercial and subsistence fishers alike increasingly adopted a range of destructive techniques that are used
still. They cast fine-mesh nets that can capture juvenile fish before they breed.
At night, after luring schools of fish with lights, they set off dynamite, whose
underwater blasts send stunned fish floating to the surface while also killing
larvae, juveniles, and corals. In a technique called *muro ami*, drive-in net fishing, fishers—often just boys—dive down and strike the corals with rocks, scar-
ing the elusive reef fish into nets and damaging the corals. Around 1990, the
growing demand from pricey restaurants for live fish brought an increase in the
use of poison. Divers squirt sodium cyanide from a bottle, which immobilizes
the large, lucrative fish, but also poisons the surrounding corals. A study of cyanide fishing estimates that since the 1960s, divers have deposited more than 1
million pounds of sodium cyanide in the Philippine reefs.[7]

As the political and economic climate deteriorated in the 1970s and 1980s,
fishing became more important to the poor. Until then, the coasts had just sup-
plemented a family's income. The 1957 fisheries journal reported, "No matter
how poor a family is, yet the head [of the family] has a piece of land which he
can call his own. Most of the people are engaged in both fishing and farming
while . . . very few are devoting themselves entirely to the fishing industry."[8] By
the 1970s, many relied more on fishing and less on subsistence farming. The
population was growing, but without effective agrarian reform, land remained
concentrated in large plantations. Military conflicts had also heightened in
rural provinces. Many farmers seeking land or fleeing the conflicts migrated to
the coasts and became fishermen. The growing competition among the fishers
set off the vicious cycle of destructive fishing. As fisheries declined in one part
of the country, fishermen either boosted their catches illegally or fished elsewhere, thereby putting more pressure on the fisheries.

As the state of the reefs and fisheries became critical, a number of groups
stepped in to try to manage and conserve the coastal resources. Government
agencies played a role, as did grassroots organizations and nongovernmental organizations (NGOs), international lenders, and academic and research

institutions such as the Marine Science Institute at the University of the Philippines and the International Center for Living Aquatic Resources Management, an international fisheries research center then based in Manila. In the Visayas, the key player was the marine laboratory at Silliman University in Dumaguete.

Silliman University, founded by American Protestant missionaries in 1901 as an elementary school for boys, is sometimes called the American university of the Philippines. It was named after a Christian philanthropist from New York, and its first three presidents were American, as were many of the faculty and staff. Silliman is known for its enduring ties with the United States and for its marine laboratory. In the mid-1990s, shortly before a modern laboratory was built with funding from the U.S. Agency for International Development (USAID), it was still a modest facility with single-story buildings and a string of saltwater tanks that sat along the waterfront some distance from the main campus and its elegant rows of acacias.

The marine laboratory has conducted decades of research on how marine reserves might benefit fisheries. In the 1970s, Angel Alcala and Alan White began to set up experimental marine reserves on several small islands near Dumaguete. Their main hypothesis—at the time unsupported by data—was that even a small no-take zone could help restock nearby fisheries. These reserves, which served initially as study sites, eventually became the basis for a whole new way of thinking about coastal management that came to be used worldwide.

The first project was on Sumilon Island, a 57-acre reef-fringed island close to the southeastern tip of Cebu. Although about 100 fishermen from other islands fished near Sumilon, only four families lived there. In 1974, Silliman set aside about one-fourth of the coastline as a marine sanctuary where fishing and other activities were prohibited. The fishers could work along the rest of the reef but could not use destructive techniques. A fisherman who served as the reserve's caretaker recorded the types of fish caught, their weight, the kinds of gear used, and the time fishermen spent fishing. Researchers gathered additional data.

Because of changes in the local government, the reserve was reopened to fishing, then closed, then reopened again. The back-and-forth gave the scientists a unique chance to study the reserve, which ended up as an unusual "natural experiment." Closing the sanctuary to fishing clearly helped maintain yields in the nearby fishery; after the reserve was reopened, the initial yields from that area would be high, then drop as it was fished out. The researchers also learned that the support of the local government was necessary for a no-take zone to succeed. The final lesson came from the families living there, who worried that,

groups have since become a standard conservation practice in coastal areas. From 1984 to 1986, White worked on a USAID-supported project that did more intensive organizing and education. It was during this project, according to White, that the community grew to accept the sanctuary, which was established legally in 1985. It was slightly larger than the initial area, its corners marked by buoys and signs that identified it as a reserve.

By late afternoon, the water had grown too choppy for the ferry to return to the mainland, so Pabs and I spent the night in the guesthouse near the sanctuary, falling asleep to the waves and awakening to children playing outside the low windows. The water lapped toward shore, its low waves crossing the reef, barely visible as white-edged ripples that sank onto the sand. Although there was no rope to keep them out, boats skirted the reserve, and even the children did not run into the waves there.

A few yards from shore, the corals in the reserve thrived, with schools of fish zigzagging back and forth in the shallow water, sunlight glinting off their scales. Near fishing villages like this, a look underwater usually reveals the aftereffects of cyanide, dynamite blasts, and blows of *muro ami*. What had been reefs look like open graveyards: beds of gray bits of coral, scattered like bleached bones on the ocean floor, without any color or movement. At Apo, a few bare spots in the reef showed where anchors were dropped carelessly, but the great coral heads covered the rest of the sandy ocean floor, colorful, undulating, and full of life.

According to the fishermen, the reserve has been a success. In surveys, fishermen claimed that their yields have at least doubled or tripled, and that the gains have more than compensated for the smaller fishing grounds. Suan reflected, "Before the reserve was set up, we could get a lot of things from that area, like sea cucumbers and small fish. Afterwards, we couldn't take anything at all. But after the reserve, the corals were more beautiful, and there were more fish.

"If I had to compare my life from before the reserve and now, it's getting easier. And I have gotten richer. Before, we might spear three fish. Now we see whole schools. The fish breed there; there's food for them there. The small ones stay. The larger ones go farther, and we catch them. I think everyone agrees that the reserve is a good thing. We've extended it two times."

Life on Apo Island has become slightly less precarious, a result found in other villages that benefited from the coastal management programs supported by the government and foreign agencies. Instead of *nipa* houses, the fishermen live in solid concrete houses with roofs of corrugated metal that can better withstand monsoons. The cooperative and weaving associations, through which women make and sell mats, have endured. People come from other provinces and other countries to observe the sanctuary, said former *barangay* official Delmo. Still, none of his nine children lived on Apo.

"Now there are only children and old people here," he said. "The others want another life, go to other places. Four of my sons fished in a big boat that went to Palawan and Borneo. They earn more compared with me. There are lots of fish here, but when there are monsoons, like now, we can't go out for fishing. The children send funds during these times."

When asked if he would rather have other work, Delmo replied, "That's a hard question. Apo is a small island; there's no possibility that any industry would be established here. If a big resort or big industry came here, people would be influenced. Garbage would be a problem, and the cultural influence. Filipino culture is different from Western culture. People [tourists] come here and swim without clothes. We want to protect our children from those influences."

In time, the reserve itself has become even more established. In an effort to thwart developers trying to buy land, in 1994, under Alcala's tenure at the DENR and the new National Integrated Protected Areas Law, the reef was made a national seascape. In time, the sanctuary came to be managed by a protected area board that included representatives from the local government, the community, Silliman, and the DENR. The community continues to maintain the reserve, although there has been some frustration at the loss of autonomy. The community also has lost full control over the fees gathered from tourists and others, but the national protection has given the reserve greater stability, ensuring that it will survive.

❧

Long before fishing became industrialized, fishermen, politicians, and scientists began debating the resilience of the world's marine fisheries. While many countries set up schemes generations ago to manage their forests, calls for fisheries conservation have repeatedly been rebuffed with claims that marine life is infinite, any apparent declines due to natural fluctuations in populations and environmental phenomena such as climate change.

In early eighteenth-century England, for example, laws were passed to regulate fisheries—out of concern that "the breed and fry of sea fish has been greatly prejudiced and destroyed." They were later overturned. Nearly two centuries later, the English biologist T. H. Huxley gave an enthusiastic lecture before the National Fishery Exhibition in Norwich that was innocuously titled "The Herring." Immediately published in the journal *Nature*, the talk began with the natural history of the small fish about which Huxley was an authority. He offered fastidious details of the herring's anatomy, reproduction, and feeding behavior. He described its food of choice, minute crustacea that "tenant . . . the ocean in such prodigious masses" that they discolored the water for miles. This immense fertility of the underwater world, Huxley then argued, meant that regulations were unnecessary and the industry should remain unfettered. With "not a particle of evidence that anything man does has an appreciable influence on the stock of herrings," the wisest course, he insisted, was to "let the people fish how they like, as they like, and when they like."[10]

Despite Huxley's pronouncements, the House of Commons concluded a decade later that some fish stocks had been "definitely impoverished," and by then even the fishing industry was convinced. By the late 1930s, E. S. Russell, the director of fishery investigations at the Ministry of Agriculture and Fisheries, pleaded for an end to the "wasteful and uneconomic" exploitation of Britain's fisheries. In the early 1950s in the United States, however, a survey from North Carolina pronounced that marine fisheries' yields actually increased as human populations swelled. "No single species so far as we know has ever become extinct, and no regional fishery in the world has ever been exhausted."[11]

By the 1990s, the debate appeared over. Today, from the coasts of South Africa to the North Sea to the Northeast Atlantic to Southeast Asia, important fisheries have been depleted, and most commercial species have been fished to capacity. Heightened fishing effort still masks the dwindling catches, but the Food and Agriculture Organization estimates that the populations of more than one-third of the top 200 fish species are already in decline, making the need to control fishing effort "urgent."[12]

While the major obstacle to conservation has been opposition from the fishing industry, another difficulty lies in the very approach used to manage fisheries. The mathematical models at the basis of fisheries science assume that an optimum level of fishing, a maximum sustainable yield, can be used to calculate how much fishing can be done without harming a particular population. In practice, the concept of an optimum yield falls short. Many models are based on measurements of populations that already are overfished, and the maximum sustainable yield is rarely observed. The models were also built using data describing oceans in the temperate zone, whose ecosystems with their small number of species differ greatly from the complex ones found in the tropics.[13]

In addition, managing fisheries requires a solid base of information—sound science about the life cycles of the fish and corals, the size of fish stocks, the interactions of different marine species. Despite the importance of fisheries, especially in Asia, much of this information is either poor or nonexistent. Given the size and depth of the oceans and the natural fluctuations in marine populations, it is difficult to gauge fisheries' size over time. Because current estimates of fish stocks rely so heavily on limited data, one group of Filipino scientists called them "snapshots at best."[14] Like the "snapshots" of fish stocks, the available science barely provides a foundation for modeling tropical fisheries.

The life cycles of most reef fish are divided into two stages, planktonic and sedentary. When females spawn, they release millions of eggs onto the ocean floor. During the planktonic stage, eggs and larvae float in the open water, drifting for as long as several months before they join a reef population and settle for the rest of their lives on the ocean floor, a process biologists call recruitment. Most larvae fail to reach this sedentary stage, but those that do live from a few years to 15 or more. As females grow, the number of eggs they produce increases exponentially; a single 24-inch red snapper may produce as many eggs as more than 200 red snappers that are 16 inches long.[15] Catching the largest fish—the breeder—is self-defeating, akin to farmers in a famine cooking their last kernels of seed grain.

Despite its importance to fisheries, little is known about the planktonic stage of life, and eggs and larvae have not been identified for most fish, including major commercial species. One scientific paper candidly calls the larvae "difficult to deal with."[16] What is known is that their survival varies dramatically and depends on the ocean currents, weather, and the presence of predators and food. But no one knows how far larvae travel and why, what proportion survives, and what makes them thrive. Their movement seems to be influenced by storms and waves, by sediment, by the wind, and by their microhabitats. Research suggests that some reef populations do replenish themselves, but others depend on larvae from afar. Larvae also are not, as was once believed, entirely passive, drifting about on ocean currents like pollen in the wind. They can exert a measure of will over where they settle, and some larvae actually swim actively. Recent studies show a lunar influence; larvae may settle on the reefs in pulses—great collective bursts—near the new moon of each month.[17]

When the Silliman researchers first set up marine reserves in the 1970s, they had to work around the gaps in scientific knowledge. There were no sure studies to guide them in choosing the size, location, or type of reef for the reserves. There was not even a consensus that sanctuaries had a measurable effect. They did not know if the reserves would help restock fishing grounds at all—or to an economically useful extent. Their hunch (later reinforced by fishers) was that small protected areas would help revive nearby depleted fisheries by providing eggs and larvae as well as adult fish that, after maturing in the reserve, would migrate to neighboring waters.

Motivated partly by the poverty in the villages, the lack of other work, and the speed with which the reefs and fisheries were declining, the researchers did not wait for scientific certainty. Even if government programs could be developed to control fishing, with so many fishermen surviving catch to catch, in villages up and down the coasts families might starve. They opted to risk that the reserves might work, hoping to develop effective ways that communities could manage their fisheries. More punitive measures that would bar subsistence fishermen from working, Garry Russ and Angel Alcala wrote tersely in 1996, were "difficult to justify socially."[18]

After setting up the reserve, they tested their prediction that the reserve would "export" fish to waters nearby. In the 1980s and 1990s, they doggedly studied Sumilon and Apo, returning over and over to gather data about the local catches and populations of fish both in and outside the reserve. Their stacks of published scientific papers have contributed steadily to the debate over the value of marine reserves.

Like all mobile populations, schools of fish are an ever-shifting mark. In the Philippines, most data come from fishermen's catches, which can mislead. A declining fishery where fishermen worked hard might yield larger catches than a flourishing fishery where fishermen had other sources of income or less efficient gear. Some fishermen also are more skillful, diligent, or just lucky. In an attempt to even out these differences, catches are measured as kilos per hour—catch per unit effort—which reflects the actual yields and the time needed to gather them. However, even this measurement does not capture the health of the fishery; with increased effort (and declining catch per unit effort), fishermen can maintain their catches while depleting fish stocks, as is happening throughout the world today.

In repeated studies of Apo, the researchers complemented the data from the fishermen's catches by counting live fish. Alcala and Russ, a marine biologist at the James Cook University in Queensland, Australia, used what are called visual censuses. Russ systematically swam up and down 50-meter lines, marking the numbers and types of selected species on waterproof sheets. Between 1983 and 1993, he took seven censuses at the reserve and at a similar reef on Apo's other side, counting a total of 178 species of reef fishes in 18 different families.[19]

The studies on Sumilon and Apo eventually showed the sanctuaries to be effective. On Sumilon, research by Alcala and Russ and others also found that, compared with a comparable site across the island, the sanctuary had a higher number of species (greater diversity) and overall higher numbers of fish (abundance).[20] The various studies on Apo also show that, although it took close to five years, the density and number of species increased. One study says that fish abundance in the reserve was twice as high as before. On both islands, the numbers of fish in areas adjacent to the reserves increased. While a natural shift in fish populations could have caused the increase, the areas closer to the no-take zones had more fish. The reserves, the authors concluded, had a spillover effect; as promised, they did export adult fish.

. . .

Conventional solutions offered for conserving natural resources typically have championed either protecting them as private property or defending them from being used with a system of centralized management. In the Philippines, coastal management had some of the same shortcomings as management of the forests. There were too many government agencies involved, and the laws were inadequate and not easily enforced. Some laws were also unreasonable or, more likely, merely symbolic; one even imposed the death penalty for dynamite fishing. While landowners could—and did—protect forestlands with private armies, guarding fishing grounds was less likely to succeed.[21] It also was unconscionable to cut off subsistence fishermen from their main source of food and income.

Alan White and others believed that a better approach was to restrict access to fisheries while turning their management over to the local residents. They argued that the coasts should remain open to a community—rather than, say, a small number of owners—so subsistence fishers could continue to work. However, in order to stem the fisheries' decline, it was crucial to limit access—by restricting the number of fishers, the type of gear used, the seasons when they fished, or the area where they fished. A small no-take zone, said White, can be effective because it affects everyone equally: "Everyone knows if it's a sanctuary. The word spreads real fast."[22]

A single island community such as Apo, when convinced that conservation has merits, can learn to monitor its own protected area. Applying this approach on a larger scale, such as to a coast or bay, was the next step, and that is exactly what Alan White and others tried to do beginning in 1996. White, who first went to the Philippines with the U.S. Peace Corps in 1978, is tall and angular. He appears to be both assertive and restrained, paired traits not uncommon in Americans who have blended aspects of Filipino culture with their own. In the mid-1990s, he joined the Coastal Resource Management Project (CRMP), a technical assistance program funded by USAID. In the Visayas, CRMP set up office in the North Reclamation Area in Cebu City, a relatively new area of landfill near the coast. Because of the importance of their fisheries and history of coastal management, the Visayas were to be the project's main focus. At first, White was one of the only staff members, and the modern building that housed the offices stood alone across from a Shoe Mart mall. Within a few years, the reclamation area would sprout many buildings, and the CRMP offices would become a bustling headquarters.

looked by foothills thick in places with palms and banana trees. The six towns lining the bay—Hagonoy, Padada, Sulop, Malalag, Digos, and Santa Maria—together cover nearly 70 miles of shoreline. About 56,000 people lived there in 24 coastal *barangays*, including nearly 8,000 fishermen. To develop a coastal management program for the bay, CRMP was working with the local governments. Each town controlled its own coastal fisheries, but the municipal boundaries in the bay were unmarked; to manage the waters, the towns needed to work together.

Digos is a dusty town swelling quickly into a city. Over the sound of wind-blown coconut fronds slapping each other, the motors of the growing numbers of tricycles whined loudly on the main roads. As was often true in the provinces, though, paper money was limp and creased and smelled of the wet market. The smaller towns were poor and their economies varied, as did their coastal resources.

Malalag, a town of about 12,000, had been practicing coastal management for years. Hagonoy, with its plantations of mangoes and bananas, was largely agricultural, and many of its residents had migrated from the Visayas. Its coastal waters were overfished, especially from the use of fine-mesh nets that gathered immature fish. Padada also was agricultural, although mostly with lowland farms rather than plantations, and its fishermen did commercial fishing in the deep seas toward Malaysia and Indonesia. Only one of the villages in Sulop, the smallest municipality, lay along the coast; the main occupation there was oyster culture, and the ground-up shells were used to feed pigs, chickens, and other livestock. Farthest from Digos, Santa Maria had 35 miles of coastline and the richest coastal resources; fishing remained the main work, and almost half of all fishermen along the bay lived in Santa Maria. The town was trying to preserve its fishery by preventing boats from other provinces or countries from entering its waters.

To develop municipal plans for managing the coast, teams of government officials ventured into the communities to gather information, fieldwork that grounded them in the reality of coastal life. "The government saw quite late the importance of marine resources," said a provincial official based in Digos. "Before, I just did desk tasks. We sat in our offices and did clerical work. Now we have contact with coastal communities. I've also learned how degraded our coastal environment is. I didn't know before."

Said the official, "We met first with the *barangay* officials and then people in each sector: the women, the NGOs, the youth, the fishers. If they're not too busy, they're glad to talk, but if they wanted to go to the sea, they weren't glad we were there. They're all still very poor. They said their lives depend on fishing. We asked why they didn't improve their lives. They talked about the lack of resources. 'We barely eat,' they said. 'We have no more fishes to catch.' We explained that when the corals are destroyed, when the mangroves are gone, there will be fewer fishes. They thought that if they caught some fishes, others would still come."

Although poor, Malalag was a quiet, orderly place, its dirt roads lined neatly with stick fences and homes of woven *nipa* palm and thatched roofs. Malalag had the bay's most effective coastal management program, and it was an example, as CRMP's Courtney said, of what can happen when mayors see the ocean as theirs. Since the early 1970s, only two mayors had held office—Andres (Andy) Montejo and, since 1998, his daughter Givel Montejo Mamaril. The elder Montejo was well known in coastal management circles. During a period when fisheries were declining, and when politicians often cashed in on illegal fishing, he had tackled managing the town's coastal resources. His reputation for understanding their importance and creating a program to sustain them extended far beyond Malalag, as did knowledge of his strong political will.

. . .

When I visited Malalag, the Montejos insisted that I spend the night at their home, and to do so brought a glimpse of the family's influence. At 67 years old, Andy Montejo had a powerful bearing; people grew quiet when he entered the room. The three daughters I met were all similarly forceful. The house had been designed to accommodate a political family. One huge room filled the downstairs, its hardwood floor highly polished. Since dawn, Mrs. Montejo had been bustling—cooking dish after dish, greeting people, watering plants, pushing forward into the day. By breakfast time, grandchildren and nieces, nephews and neighbors had begun stopping by to eat, seek counsel, or pay their respects to the former mayor.

Montejo was elected in 1972, the year that Ferdinand Marcos declared martial law. Marcos banned destructive fishing, but although catches were already dropping in the major fisheries, the laws were rarely enforced. Montejo realized that Malalag's economic well-being depended on its fisheries. "In about 1975, we noticed that the catches were getting smaller and smaller," he said. "We told the fishermen to stop using dynamite, because it will kill the small fish, and to stop using poisons, because that will kill the corals, and to stop using certain gear. It took us a long time to convince them. By 1988, the majority of fisherfolk understood and had stopped using destructive methods. After that, we observed that their catches were increasing—from two kilos a day to seven to ten. And the fish were big."

His effort received support from a group of Silliman students who assessed the state of the reefs. They found that because of the illegal fishing and silt washing down from the eroding hillsides, the coral coverage had declined to about 10 percent. (By contrast, only one-third of Apo Island's corals showed damage.) Montejo pushed for an ordinance to set up a marine sanctuary, which the town eventually passed. As on Apo, it took two years for the community to accept the restrictions on fishing.

According to a number of interviews I conducted, in most towns around Malalag Bay, illegal fishers still rarely faced arrest. Some of them were too influential to be reined in by the law. Others were poachers from elsewhere. Still others were related to government officials or potential witnesses. Those violators who were detained generally were never charged, or the cases languished or were withdrawn, the fines left unpaid. As mayor, Montejo had made an effort to enforce the laws being ignored elsewhere. By the late 1980s, Malalag's eight-mile coastline could still support more than 1,600 fishermen.

"When we put up the fish sanctuary, some were against it. If we did not exercise political will," said Montejo, "people will insist on their method of fishing. So we assigned a policeman there. We caught 6 or 7, then 12, and filed cases against them. They asked that the cases be withdrawn. I said no, and many pleaded guilty. Those who were convicted before are now with us. There's no more illegal fishing, and we'll continue to rehabilitate the reef."

In Manila, discussions about the importance—and weakness—of local government inevitably turned toward the topic of political will, which too often was lacking at the local level. Montejo claimed that his stand on the coasts had not weakened him politically. "It's a long process," he said. "I tell other mayors not to worry about losing votes if they support coastal management because they'll gain more votes from people in favor of the program."

As in San Fernando and on Mount Kitanglad, the power of traditional politicians persisted, especially in rural areas. Often disparagingly called *trapos* for short—a word that also means "old rag"—they ran villages, towns, and even provinces like their own fiefdoms. Their political machinery was described as "goons, gold, and guns." Local elections, suspended during martial law, were restored in the 1980s, but even then, many provincial politicians who had held office during the Marcos years remained powerful figures. Even after a transition of power, the political dynamics in a town or province could still revolve around the major political family and seem opaque to outsiders. "Local politics is crucial," said one of the CRMP staff, an observation deemed so provocative that it was followed by a request not to be quoted by name. "Usually it's a few key people who make a program work or not. A community will still look to the

Fishing illegally. Malalag Bay, Mindanao.
RYAN ANSON

mayors to provide for them. They run the economy as the *datus* did. This is the reality at the local level."

I had expected that my key question for Montejo—why he had the vision to exercise political will about coastal management—might have a complex answer. I did not, however, expect him to avoid the question. In the evening, we spoke on his wide porch together with one of his daughters, Kinking, who worked with an NGO. A slight, sharply beautiful woman, she sat to the side when I spoke with her father, smoking cigarette after cigarette and saying little. It was quiet in Malalag. The sun had gone down, the evening mosquitoes had come out, and the visitors had trickled away. The unnerving, sweet smell of pesticides being sprayed wafted over from the banana plantation.

I pressed Montejo several times about political will, but each time he looked away, pretending not to hear. In the morning, as I sat on the porch after breakfast, he approached and answered the question abruptly, if obliquely. "We only used political will as a last resort, when all other means—democratic and otherwise—failed," he said, then walked away. It was not much of an answer, but it was all that he was willing to say. Kinking tried to bolster his answer, perhaps to update it to the current era and soften its implication that political will meant force and was undemocratic. "Political will is the willingness to enforce the law and to empower people," she said. "You have to understand the people, their income, what they need. Political will comes from that understanding." Another interpretation, said one of the CRMP staff, was that despite his reputation for protecting Malalag Bay, Montejo himself was not able to control illegal fishing.

. . .

Although *trapos* still prevailed in rural areas, since the early to mid-1990s, towns and cities had begun to elect new mayors who were young, educated, and motivated. Greeted with hope across the country, many of these young mayors were grappling with the state of their natural resources; like the provincial official in Digos who had learned about the coastal fisheries by interviewing fishers, they were more likely to understand how degraded their economic base had become. They also were trying to address the poverty that the declining fisheries brought and that in turn hastened and deepened the degradation. Elected the previous year, Givel Montejo Mamaril belonged to that new generation of politicians.

Like her father, the mayor was a fighter, and she was fighting one of Malalag's persistent problems, fishing in the adjacent waters of Santa Maria. Although the large commercial boats were supposed to be restricted to the deeper waters, they often intruded into the shallower, near-shore fisheries because the municipalities were too poor to mark their boundaries. The competition between small and large boats was growing countrywide, and advocates of the small-scale fishermen blamed overfishing on this competition—and the growing export market.

Santa Maria had voted to override the national law and allow commercial boats into its municipal waters. The town also allowed widespread illegal fishing, including with dynamite, which concerned its neighbors. "There are no boundaries, so you can't say that this is yours and this is mine," said a Malalag council member. "If people are doing illegal fishing [elsewhere in the bay] and are apprehended, they will just say that they were in Santa Maria, where they have a permit to fish."

The mayor had confronted officials in Santa Maria and tried to shame them into controlling the illegal fishing there. "The more pressing need is how the 1,600 fishermen in Malalag can continue to feed their families. I went to Santa Maria and told them, if you allow blast fishing, the damage will not stop at your area; it will extend into my area. I told them they didn't have the guts to apprehend the blast fishers."

Malalag had little economic development, which meant few jobs and few assets to attract investors who might bring jobs. Its picturesque foothills, rising abruptly from the bay, revealed the acute need for land and work. They were covered with a patchwork of *kaingin* where farmers had planted cassava, corn, *camote*, legumes, and *tanglad*. Even where the hills were nearly as steep as cliffs, they had been shorn of trees and planted almost from base to crest.

Givel Mamaril and I spoke late in the evening, after she returned from visiting remote *barangays* high in the hills. She had been widowed shortly after the election the previous year, and as we spoke on her porch, her young son and daughter joined us. The girl pressed herself into her mother's lap and fell asleep. Nearby, a large fish tank bubbled. The mayor's face was tired. She told me that she slept with her children to give them the time with her at night that they did not have during the day.

The mayor also revealed that, although working with CRMP, she was exploring other economic options for Malalag. She recently had returned from a convention for mayors of coastal municipalities, part of an effort to groom the new generation of mayors. They had recently become responsible for their municipal fisheries, but most local officials lacked the necessary technical skills to manage them, and many still did not recognize the importance of coastal management and the way it could benefit a local economy. Although the decentralization of the government directed tax revenues to help local officials support their new functions, they did not always have sufficient resources to carry them out.

Workshops at the convention addressed these needs. Some covered the financial dimensions of managing the coasts' resources, and the mayor was still digesting what she had learned about poverty in coastal areas, the enforcement of fisheries laws, ways to finance resource management. In our conversation, she spoke of green taxes and other charges that could be levied on those who used the natural resources; she was trying to find lessons that might actually work in Malalag, she said. There was a move toward "green accounting," toward including the use—especially the unsustainable use—of natural resources in calculations of national income. Counting the costs of coastal degradation could increase the internal revenue allotment—the percentage of national tax revenues that towns received—giving local governments more funds to set up coastal management programs that might, in time, pay for themselves. Although intrigued, the mayor could not easily see a solution there to Malalag's problems. While she was committed to conserving the natural resources, she was not sure that doing so alone could lift the town out of poverty and underdevelopment.

"It's too late that people realize that natural resources have value," she said. "We have to figure out how to compute them, but we don't have enough knowledge, enough data. There's a move to include coastal resources more in accounting for the IRA [internal revenue allotment]. But by the time it comes, the resources will be so depleted, what good will it do?"

Although her father had pushed for coastal management, he believed that ultimately only industrialization could create enough jobs. Like him, the mayor did not see that Malalag had a lot of economic options. While wanting to continue to support conservation, she was struggling to define a future for the town. "How can we provide alternative livelihood for these fisherfolks whose resources are already overfished? If the fishermen have to stop fishing, where would they go? It's the only skill they have."

Only toward the end of my visit did I learn what the CRMP staff learned after they set up their program. Apo might not have other options, but Malalag Bay was slated for industrialization. In the late 1980s and early 1990s, in an effort to distribute investment, the Aquino administration had created a chain of provincial agro-industrial centers, one of them along Malalag Bay. The municipalities working with CRMP had pledged about 5 percent of their land to the center. When approached by CRMP staff, they also agreed to work with them.

According to Angel (Jimmie) Ragidor, the optimistic 49-year-old officer in charge of the secretariat for Malalag Bay's agro-industrial center, the mayors had joined together in an effort to attract investors. The towns had many fruit trees, particularly mangoes and bananas, and the market for dried fruit was expanding. They hoped to bring the fruit processing there and keep more of the harvests' value locally. "We're looking for investors," he said. "Outsiders say we have many resources and should be able to raise our standard of living." Some investors had expressed interest, but the water and power were unreliable and the roads inadequate. A representative from an international aid agency had reported that "the infrastructure was bad, and no investment was coming in," he said. Still, the secretariat would persist.

An outsider might see duplicity in the local governments' willingness to accept funds for conservation while hoping to industrialize. In Malalag, that judgment seemed harsh. "People are concerned about the environment now. They realize that it anchors everything," said Jimmie. In another place, where fish were still abundant or less central economically, where landless farmers were not tilling steep cliffs, where most families did not depend on the natural resources to survive, his statement might have sounded gratuitous. There, it sounded determined. "We hope to pick industries that are environmentally friendly," he said. "If I'm pessimistic, that's the end."

The mayor was trying to be sanguine. "Would industrialization necessarily make things worse? We're trying to anticipate the development," she said. "Maybe the two could coexist, fishing and factories. There are more developed countries that have preserved their marine resources better than we have. Besides, we degraded our coastal areas *without* industrializing. If there are industries that would give more hope, more economic potential, then it is up to the people to choose. If they're poor, they probably will choose economy first."

6 | "No Need Cement"

In July 1996, a delegation of government officials from Manila arrived in the out-of-the-way town of Bolinao, Pangasinan, almost halfway up the island of Luzon. By then, Bolinao was already well known for a conflict over a proposal to build a cement plant that would abruptly transform the string of fishing villages into a sprawling industrial center. For nearly two years, the debate over the proposal had split the town. The future of the cement plant—and therefore the future of Bolinao—depended on the Department of Environment and Natural Resources (DENR). The agency could either issue the needed environmental compliance certificate (ECC) or deny it, and its decision was imminent. The DENR delegation arriving in Bolinao that afternoon included Antonio La Viña, the undersecretary responsible for the case. The time that Tony was devoting to the case, including making the long trip to tour the town and meeting with the public, underscored its importance. The situation there was charged, and Bolinao had become the highlight of national attention and public debate.[1]

The Philippines was moving into a new stage of industrialization. The Fidel Ramos administration stressed regional development: to attract investment, to facilitate trade with the booming economies in the region, and to even out the development that previously had been concentrated in Manila and Cebu.[2] The industrialization focused on industrial zones identified for each region. In these zones, often just provincial towns, change was coming quickly. In many of the towns, residents felt as though they were about to embark on an entirely new trajectory—toward traffic and traffic lights, toward cement roads and motor vehicles, toward factories and wage labor, toward all those irreversible changes that reverberate through society and might be called progress. In Bolinao, this shift was in question.

In another time and place, the construction of the plant and transformation of the town might have been inevitable. The Philippine government had echoed the call for sustainable development put forth by the 1992 United Nations Conference on the Environment and Development. The country's national plan for economic development for 1993 to 1998 described the "urgency of industrialization," an industrialization that would be sustainable so it would not burden future generations.

The plan stressed the importance of local autonomy, made more possible by the recent decentralization of the national government. Provincial and local governments could improve the management of natural resources, as with small marine protected areas, but greater local autonomy also might bring a higher level of democratic participation, which too often had been lacking. "The past is replete with lessons from grandiose schemes that have foundered on community resistance, simply because local opinion was not taken seriously," the plan continued. The goal would be "genuine consultation" at the local level.[3]

There was much discussion about whether this vision of sustainable development could actually translate into real decisions and effective policies, and Bolinao had become the test case. Democratic consultation meant asking whether or not the community wanted the plant, despite its size and economic promise. That question turned out to be difficult to answer. At the community level, part of the town supported the plant, but part opposed it. While the dilemma in Bolinao mainly focused on how the decision about the town's future would be made, it also raised the larger question of whether the Philippines should accept notoriously "dirty" industries, such as cement, that other countries refused or instead try to pursue a cleaner, more environmentally sound path of development. Finally, the newly devolved government was not yet coordinated at the national, regional, and local levels. In Bolinao, as elsewhere, development policies conflicted, agencies' responsibilities overlapped, and there was minimal land-use planning. Given all this uncertainty, the real question in Bolinao had become who should make the decision.

As DENR undersecretary, Tony La Viña was in the middle of the debates and confusion. A human rights and environmental lawyer, he had spent a number of years working in NGOs before being appointed undersecretary. He was part of an influx of young government officials, many of them from NGOs, expected to be dedicated more to public service and less to their own enrichment than officials molded by the Marcos era. In an agency known for corruption, Tony was hoping that the Bolinao case could set a precedent for future decisions.

Heading north.
Northern Luzon.
MARISSA ROTH

Late one July afternoon, I joined Tony and his staff on their way to Bolinao. The Usec, as his staff called him, is a handsome man who exudes intelligence. He sat in the front seat, simultane-ously instructing the driver in Tagalog, answering my questions in English, and talking to his staff in both languages. For the first hour or two, we made good time, moving steadily north on the concrete road.

At first the air-conditioned van insulated us from the heat and poverty and made it easy to forget the conveniences left behind in Manila. Even in the mid-1990s, though, it was not possible to travel through Central Luzon and remain unaware that the violent eruption of Mount Pinatubo had transformed the region. The largest flat area in the Philippines, the rich plains of Central Luzon had been its rice basket. Across the provinces of Pampanga, Tarlac, and Nueva Ecija, flooded fields had stretched in all directions, and the rice grown there had made the region a prosperous place dotted with two-story wooden homes. Each season brought to the landscape field laborers bent over in dif-ferent tasks and new hues of green plants, lengthening in unison, that rippled in the wind.

Mount Pinatubo's initial eruptions in 1991 deposited several million tons of volcanic debris that covered mountains, crushed buildings, and collapsed bridges. What followed, although little known outside the country, proved more destructive. Year after year, the seasonal rains beat down, mixing the sand, gravel, and boulders that blanketed the peaks into a slurry that ran down the hills toward the ocean. As riverbeds filled with volcanic debris, mudflows of *lahar*—the Indonesian word for this liquefied ash—forged new pathways into the valleys, spreading out over the plains. The color and consistency of wet cement, *lahar* can flow quickly enough to overtake a running man. It never dries fully, and with each heavy rain, it absorbed water and surged ahead, downing bridges and covering up villages. The spreading *lahar* had turned whole parts of Central Luzon into new desert punctuated only by occasional roofs and tree-tops poking through the rough ash. In some towns a few families, rather than abandoning their homes, raised them on high stilts, transplanting fruit trees into the hot, gray wasteland and futilely closing their windows against the dust and grit.

Five years after the eruption, the *lahar* zone intruded through large parts of Central Luzon, following the valleys and rivers. For road travelers, it remained a bleak bottleneck. Trips between Manila and the northern provinces that once had taken three hours required an unpredictable five or six. Driving through the *lahar* zone was like driving through a lingering sandstorm. In jeepneys and other open vehicles, the airborne sand reddened eyes, coated skin and hair, dried mouths, and settled scratchily in one's lungs. Even in an air-conditioned vehicle, the air would grow thick.

When we reached Pampanga, two provinces away from Pangasinan, traffic slowed, and then in Tarlac, not far from Pinatubo itself, it slowed further. Deep in conversation, the undersecretary looked up in apparent surprise. It's the *lahar*, someone offered. There was no more to say.

After one trip through the devastated parts of Central Luzon, the remaining trips seemed the same. They began with the slowing of traffic, the delays in the middle of blackness. There would be a few lights up ahead, single lights low to the ground that jerked around, the lights of flashlights or torches. As the vehicle inched forward, there would be shouting in the dark, as boys or young men came into view, the contours of their faces almost visible as they halted the vehicles or waved them forward, their eyes shadowed in the uneven light, their heads swathed with towels, their noses and mouths covered with handkerchiefs as they stood calf- or thigh-deep in muck. There was always the same dependence on them to guide the vehicle, the same wondering about who they were and what they themselves had lost to Pinatubo. There was the same blackness; the worry about the vehicle; the laboring movements over uneven ground; the eerie, sloshing echoes beneath the undercarriage; the barriers of sand-filled rice sacks that the *lahar* would eventually overwhelm.

By night, the government van had reached Dagupan, Pangasinan. The *lahar* was far behind us. Through the darkened windows, I could see evidence of the new wave of economic growth. The roadsides that a few years earlier had been dark now were lit by the bright signs of hotels and fast-food restaurants. They were also a sign of the changes being seen across the country. In the provincial areas slated for industrial development, including Pangasinan, buildings, bridges, and roads were being constructed. With them came a growing need for building materials, particularly cement, and particularly cement that did not have to be trucked through Central Luzon. Along the Lingayen Gulf in Pangasinan, the deep, high-quality coastal deposits of limestone, a key ingredient of cement, looked promising.

⚡

Pangasinan is a crescent-shaped province on the western edge of Northern Luzon that is flanked by the Zambales Range and the Cordillera Mountains. The province is often described as a gateway, both to Baguio and the north and the rice fields of Tarlac and the south. The province also cups the Lingayen Gulf, a wide, deep channel that opens to the South China Sea and shipping routes to Asia.

Pangasinan is one of four provinces that make up northwestern Luzon, a coastal region that, according to the main business newspaper, despite the region's "natural attributes" remains "raw and left out." In 1996, about half its residents lived below the official poverty line, and about 96 percent did not have enough work. The provinces planned (although the planning was more in the hoping stage) to develop ecotourism, particularly in the Lingayen Gulf. As along Malalag Bay, however, plans conflicted; in 1992, northwestern Luzon had been identified as a new industrial zone, the North Luzon Growth Quadrangle (Northquad). With its proximity to Taiwan, Northquad would be the "springboard into the Asian tiger economies."[4] Pangasinan, the birthplace of then president Fidel Ramos, was to be a model growth area. The first step toward Northquad's economic makeover would be the cement complex in Bolinao.

At the farthest end of Pangasinan, at the edge of the Lingayen Gulf, Cape Bolinao juts like a fat thumb into the South China Sea, pointing northward to Hong Kong, China, and Taiwan. At the tip of the peninsular knob sits the town of Bolinao, named after the Tagalog word for herring. The second largest of the towns along the gulf, it is made up of 20 *barangays*, most of them coastal, with nearly 10,000 families in 1992, or more than 52,000 people. There was little agriculture, mainly fruit trees and cattle; the major industry was fishing, on which nearly one-third of the families depended. The subsistence fishermen barely eked out a living, and as the fishing degraded the coast further, their economic future grew bleaker.

The town of Bolinao dated to the Spanish era, its center the traditional square bounded on one end by a church with barred windows and stone walls streaked black with lichen. In the evening, a sole window shone near the church's roof. Most of the shops closed by seven o'clock, but long after sunset a line of vendors continued to sell hot dogs and pork barbecue from small tables. Hands exchanged meat for money in the light cast by oil lamps, the small yellow flames each a smear that barely pierced the dark. The sky was black and, to those just arrived from Manila, full of stars. The town awakened early, and by sunrise, uniformed children walked to school and lines of tricycles and jeepneys waited for riders. Along the steep and winding road leading down to the shore, tricycles maneuvered one by one over the rutted mud, at the most precarious turn each waiting until the one in front had slid to safety before moving ahead.

On the coast, not far from where the road straightened out, stood the Bolinao Marine Laboratory of the Marine Science Institute of the University of the Philippines, in Bolinao called simply MSI. Its main research building abutted the water with a view of the island and cove so fetching that, said an older resident, the area should have been developed as a resort instead. Scientists at MSI had been doing research there since the mid-1980s, and in the early 1990s, they had gotten involved in organizing the community and helping develop conservation programs. In the controversy that developed over the proposed cement plant, the presence of MSI proved important, as did the quietness of the town and its long history, the large number of people living along the coast and their dependence on fishing. In retrospect, it seemed surprising that the corporation seeking to churn out cement in Bolinao could have so underestimated the objections that would be raised.

In the mid-1990s, the cement industry in Southeast Asia was growing quickly. Construction in the region was booming, and dozens of plants were being built to supply it. In the Philippines alone, the industry's capacity quadrupled from the 1980s to the 1990s, with most of the expansion taking place in just a few years. Between 1993 and 1996, the DENR approved about 25 projects related to quarrying and cement production. Only three were in Northern Luzon and one in Pangasinan.[5]

In 1994, with the Philippines still apparently facing shortfalls of cement, the Pangasinan Cement Corporation, a consortium of companies from Taiwan, Japan, and the Philippines, proposed building a 525-million-dollar cement complex in Bolinao. With a power plant, cement factory, wharf, and sites for 22 limestone quarries all linked by 6 miles of covered conveyor belts, the complex eventually would spread over parts of five *barangays* and almost one-third of the town. It was expected to provide cement for the Philippines and other parts of Asia—it would be equipped to ship bulk cement to Taiwan—and to jump-start industrialization in Northquad. The consortium's promotional brochure promised that the proposed complex would help transform the country into "an emerging economic force in the Asia Pacific."

The company also presented the facility as one that would usher in greater efficiency and better environmental measures for the growing cement industry. As a "world-class cement plant," read the brochure, the complex could, like its sister plant in Japan, coexist productively with fishing, farming, and tourism. "[A]s long as the state-of-the-art anti-pollution devices are installed and used continuously, and the recommended safety measures [are] . . . faithfully adhered to, there shall be no objectionable impact to the environment," particularly the Lingayen Gulf. Most existing plants in the Philippines used out-dated technology, and the presence of a more modern one would spur others to "update their processes and equipment to remain competitive" and thereby help raise the industry standards.

When the plant was first proposed, it met with a strong and divided reaction. Some citizens, fearing the instant metamorphosis from fishing villages to factory town, formed the Movement of Bolinao Concerned Citizens to mobilize against the project. The *sangguniang bayan*, municipal council, came out in favor of the plant. *Barangay* officials signed petitions on both sides, and the mayor fluctuated. MSI, located a few hundred meters from the proposed site for the plant, worked with local citizens' groups that opposed the project. Faculty of the University of the Philippines lobbied hard against it, and scientists from around the world sent letters of objection.

While the quarry alone would replace about 200 families and affect at least 3,200 more, the main concerns focused on the health and environmental risks. Construction of the wharves would require dredging the ocean floor, and once in operation, the plant would pollute both air and water. The quarrying operations would cause erosion and siltation, which would further erode the reef. The complex was expected to need hundreds of tons of fresh water an hour, near a town center that did not use three tons per hour. It also would need for cooling about 1,000 tons of seawater that would be returned to a coastal area valued for having the gulf's most intact reef. Overall, the development was likely to destroy the natural resources that supported thousands of families. Bolinao's "land, coastal and marine resources are the lifeblood of its people and economy," said a letter signed by university board members and faculty, faulting the project as one that would "introduce an extractive industry into a natural-resource based municipality."[6]

A response from the Pangasinan Cement Corporation claimed that because the complex would be so much more advanced than plants currently operating, comparisons with them were "quite unfair and unfounded." The proposed complex was being "criticized for acts or omissions of old and pollutive cement plants." Besides, it suggested, other types of development in the town were unlikely; while plans might promote ecotourism, there were as yet no investors, and the infrastructure—roads, telephones, and other basic services—could not support a swell in tourists.

Late in 1995, after a year of debate, the DENR denied the project its necessary certificate of environmental clearance because of the "yet serious and unmitigated problems." There were, the letter said, "technical issues related to pollution," and there was a "land use conflict" over the northwestern part of the Lingayen Gulf. There also was a "lack of social acceptability" based on the democratic consultation, which the development plan had identified as important to the process of sustainable development. "There is still strong opposition to the project," read the letter; the community had objections that had not been addressed.[7] The lack of social acceptability was critical.

Over the next months, the cement corporation submitted a revised proposal. The DENR added new members to the committee reviewing the project—specialists in marine biology, hydrology, and geological sciences. By July 1996, as the deadline for the committee's new decision neared, the debate grew increasingly contentious. Beyond Bolinao, from the provincial to the national level, more voices had joined the fray: the governor (opposed), regional officials (opposed), a congressman (in favor). The powerful Speaker of the House of Representatives criticized as "fake patriots" faculty at the University of the Philippines who organized against the plant. Stung, they in turn sent the press documents showing that the Speaker's family owned the site for the plant. In Bolinao, divisions deepened even further.

Bolinao might have received less attention if not for an environmental calamity on the small island of Marinduque a few months earlier. Marinduque, which lies about 100 miles south of Manila, had been the site of a large copper mine managed by the Canadian mining company Placer Dome. For decades, waste from the mine had been dumped in a drainage tunnel, which in March 1996 collapsed without warning. More than 3 million tons of dense mine tailings "burst forth," according to one account. The spill "choked off all life" in the 16-mile Boac River, one of the island's two major rivers. Although no one died, the river flooded, covering croplands and stranding thousands of villagers. The sulfur-colored tailings contaminated the river channel all the way to its coastal mouth. The island was declared a calamity zone.[8]

Images of the Boac River clogged with orange mine waste made clear to Filipinos who might never have seen a smokestack or effluent pipe the environmental consequences that industrialization could bring. Marinduque gave a name to fear. What might previously have been a vague unease about the unknown became a more specific worry. The disaster also put pressure on the DENR to be more careful in awarding its certificates of environmental compliance. The agency's next major decision was over Bolinao.

❧

In the early 1900s, when the Philippines stirred endless fascination in the United States, Emma Helen Blair and James Alexander Robertson translated from Spanish 55 volumes of fastidious records kept over hundreds of years by navigators, missionaries, and functionaries. Like Dean Worcester's books, the accounts of the islands are detailed but slanted. They do, however, give a feeling for day-to-day life and for the origins of the rural villages. According to a 1590 account, *barangay*, the old Tagalog word for tribal communities, originally meant "boat." Individual *barangays*—as many as 100 homes—consisted of people who had arrived on one boat. They were only loosely clustered into towns. "This barangay was a family of parents and children, relations and slaves. There were many of these barangays in each town, or, at least, on account of war, they did not settle far from one another. They were not, however, subject to one another, except in friendship and relationship."[9]

Today, the *barangay* remains the smallest political unit, although it in turn is made up of individual *sitios*, or communities. A *sitio* can still reveal a person's identity. In a rural area, people asked where they live will begin with their immediate cluster of homes—the *sitio*—and then name the *barangay* and municipality. If people know what *sitio* you are from, they know your family, a young fish warden from the coast of Mindanao told me, and if they know your *sitio* and your family, they know who you are.

On an earlier trip to Bolinao, I had learned the importance there of knowing who you are and where you are from. I had taken a midnight bus with an oceanographer and his research assistants, bouncing north along the same slow road that I would take seven months later with Tony. The conflict splitting Bolinao was obvious. Banners for and against the proposed plant hung from windows. "NO NEED CEMENT" had been painted in large letters on a bridge overlooking a brackish inlet where coconut husks lay heaped in high piles on the shore. The question of what would happen to Bolinao came up in most conversations: when we had beer in the two-story restaurant built around a thick, gnarled tree; when we chatted with the grandmother next door; when we talked with people in outlying *barangays*.

On Sunday, the oceanographer's assistants delivered me to Josue Aragon, the Methodist pastor in the village of Balingasay since 1978. Probably the most educated man in the village, the pastor would undoubtedly want to talk with a visitor about Bolinao's decision and its future. We headed down the narrow beach, away from the knot of homes where he lived with his wife and three children in a *nipa* house on stilts. It was high tide, and the waves washed onto the sand near our feet. In his late 30s, the pastor considered Bolinao his home, and he was open about how townspeople had responded to the proposed cement complex.

They were all grappling, he said, with the kind of economic development they wanted. Those promoting the plant "only present one side of the coin: employment, money, infrastructure," he said. But they did not talk about what those changes could mean. "There will be a lot of people striving to get the jobs. There will be a lot of businesses, bigger stores, and nightclubs. We'd have employment perhaps; we'd have money," he said. But there was no way to know for sure what they would lose, and he worried about that. "If you overwater a plant, it will soak up too much. It's like that with development," he said.

It is easy for an outsider to look at a community such as Balingasay and see only its beauty. Tourists travel half a world for this: the blue line of the horizon, the open sand, the palms, the simple homes. Once when I praised a landscape of hills to a Filipino friend, he stopped me, correcting my failure to see them fully. There's nothing there, he said, nowhere for people to work, nothing for people to live on. That's what makes them so beautiful.

The pastor, however, was not an outsider. He valued Bolinao the way it was, not for the beauty, but for the life there; he valued it because it was his. "They're talking about putting the plant here, whether we want it or not. Do we want this plant? That should be the question, instead of it being pushed on us. The development will bring factories that will pollute our air and water. Where will their waste go? Into the sea. And what will happen then? The corals will be destroyed. A few hundred people may benefit from the jobs, but thousands of fishermen will lose the privilege to fish." He seemed to mean "opportunity," but the word he used was *privilege*: "Thousands of fishermen will lose the privilege to fish. What will happen to their children? What will they eat?" He also was concerned about the cultural influences that the development might bring. American television and movies have infused cities and villages around the world, but, the pastor made clear, they were not always welcomed, even in a country dominated by colonial powers for hundreds of years.

The beach was lined with long boats painted red and blue and yellow with matching outriggers and gunwales. The houses, built close to the shore, were hidden by lines of drying laundry. Sunday was family day, and as we walked, we passed groups of all ages gathered on the woven floors of gazebos and boys and adolescents lounging on the boats, as in another setting they might have gathered around a pickup truck or car. The pastor waved the back of a hand toward them. What will be lost is this, he said, a kind of life that revolves around families, a life where people have time for each other, and where having time for each other matters. That culture and those values would change. "Do you see how we are already losing our culture?" he asked. "We always adapt to foreigners. It makes me very sad."

We left the houses and boats behind and walked beside a line of squat coconut palms leaning toward the water. The pastor, skeptical about the cement corporation's claims, brought up another concern. "The question should be asked, 'Do we have the right technical skill to handle this situation?' I understand the Filipino character. We have a lot of complacency. If a reactor is broken, the manager will pay the engineer not to speak up. 'The people aren't complaining,' he might claim. 'It's better not to say anything.'" So if a problem comes up, no one will mention it, he said; "That's what I'm afraid of."

In spirited academic conversations in the United States, risk shrinks to an abstraction, something that people with choices can accept, or not, something that benefits could eventually balance out. Risks and benefits then become compressed further into an equation that disinterested observers might use to force a decision. Here in Bolinao, where choices were fewer and life was lived closer to the margins, instead of being chosen, risks were more likely imposed from the outside. The benefits themselves, rather than balancing out the risks, probably accrued to someone else. The pastor understood the equation well. "We don't need their money if it harms our health and environment," he said. "We do need their money, but not at that cost."

At the end of the beach, as we turned back, the pastor also reversed the direction of the conversation, asking if he could ask me some questions. My guides in rural areas often made this request, although typically after several days. The questions usually were personal, about marriage and belief, and sometimes political, about the United States or about how regimes in other countries I had visited treated their poor.

The pastor's questions came quickly, as though stacked up in his mind. He started with personal questions and broadened to queries about the United States and its recent recession, then raised a few metaphysical ones. Do you think it is possible to be all good or all evil? Do you agree that the source of the life force is the tension between good and evil? He then circled back to our earlier conversation, and this line of questioning gave me a deeper understanding of how people living in a coastal village might imagine the more modern world that they had only glimpsed in newsreels or movies. The pastor wanted to know what I, coming from an industrialized country, might predict for Bolinao. I had grown up around factories; they were real places to me, not unknown specters. He wanted to know what living in an industrialized society might be like. Was it something to fear, or just another kind of life? "What will happen if the plant is built?" he asked. "What will be lost? What do you think we should do?"

There were no answers to his questions. Instead, I told him about some of the history of the Great Lakes and the cities that had formed on their shores, and about my personal history of growing up in Cleveland, Ohio, and how that had influenced my views of environmentalism. Lake Erie was still fouled with pollution when I was a child, and I thought that dead fish littered all lake shores and that all lake water smelled of rotting flesh. The environmental laws enacted in the late 1960s and early 1970s quickly had a noticeable effect, and by my eighteenth birthday, the lake was clean enough to swim in. I told him about the Cuyahoga River, which runs through the city, its broad floodplains, the Flats, lined with vast factories. In the 1960s, their waste had so saturated the Cuyahoga that more than once the oily slick on its surface caught fire, bringing the city ridicule for having a river that burned. In time the Cuyahoga, too, was somewhat revived, and nightclubs were built along its banks in the shadows of the bridges and smokestacks. When environmental laws are enforced, industrial waste can be cleaned up, I told the pastor. I held back from sounding either too optimistic or too pessimistic. This is what *can* happen, but it's expensive, and it takes political will. Maybe this is what can happen in a rich country. It did not mean the same thing would happen there, and it did not mean that the Philippines should take the same path.

We neared the pastor's home. One of his young daughters, just learning to drive a motor scooter, was circling in small loops beside the stilt house. Her eyes shone with confidence. On Mount Kitanglad and in remote villages, I had met people who had never driven a motorized vehicle. Here, no one seemed concerned; she might have been chasing a hen or turning cartwheels. Before the young researchers came to collect me, the pastor poured Coca-Cola from a glass liter bottle into coffee cups, and we drank it, squatting on our heels beneath a tree outside his house.

He had kept a request until the end. "Many writers come here and portray the Philippines in a negative light." The stories that westerners told about his country and many other out-of-the-way places too often focused on the lurid

Fisherman's house. Lingayen, Pangasinan.
MARISSA ROTH

and chaotic, on the despair and hopelessness. He phrased what he said next politely, although he spoke directly. "Please don't do that," he said. "You have seen enough of our country not to need to do that."

⚘

In the early 1960s, with one of the strongest economies in East Asia, the Philippines seemed certain to develop faster than the rest of the region. Instead, as its neighbors bounded ahead with decades of unprecedented growth, it lagged behind. Analyzing why the country did not prosper continues to occupy economists, historians, and journalists alike, with each historical period bringing a new explanation: The centuries of colonialism fostered dependency. The islands are fragmented without a sense of unity. The allegiance to the family as society's central unit further obstructs national unity. The Marcos dictatorship and martial law destroyed the economy. The small elite acts only to enrich itself. The weak government—marked by an unreliable bureaucracy—in turn has favored the wealthy and curtailed economic development, impoverishing much of the population.[10]

By the mid-1990s, the weakness of the government was drawing particular attention. The economy was growing, the United States had withdrawn from

and the next generation to enjoy. It is our sacred duty to protect, conserve, develop and manage these resources." The gulf, another said, is "God's gift, our heritage, our responsibility." During the meeting, the commission's executive director, Valerio Perez, spoke with great seriousness, as though burdened by responsibility for the body of water. He was an older man, thin and intense, and I was told that when he presented his position at an earlier meeting with the DENR, he nearly cried. There was no single plan for the Lingayen Gulf, he said. Although it had been designated an environmentally critical area, the various government bodies involved had conflicting opinions and policies. "There are those who would like tourism to come in, and there are those who would like nonpolluting industries to come in," he said. Plans for Bolinao and the Lingayen Gulf created at the local, provincial, and national levels did not always mesh, and no one knew whose authority counted the most, that of the mayors, the commission, the provincial agencies, the regional bodies, or the national DENR. "If the commission has come up with a land-use plan, would an environmental compliance certificate override that? Who," Perez asked, appealing to the undersecretary, "can make the final decision?"

Tony did not take long to answer. Bolinao is a test case, he said. Where jurisdictions and land-use plans overlapped, it was not clear who had the right to define the land use of a place. He had come to Bolinao because he believed that a town so deeply tied to its own natural resources should have a say in the future of those resources and its own economic future. The decision over the environmental compliance certificate for the proposed cement complex should be socially acceptable to the community. But no one was quite sure what social acceptability would mean.

. . .

When the undersecretary's van pulled into the town center that morning, several hundred people filled the plaza in front of the church. Umbrellas mushroomed over the crowd, as did hand-lettered signs in both Tagalog and English: "No to Bolinao Cement." "No to ECC." "R.I.P. Bolinao 1575–1996." The crowd did not yell or cheer. They just watched and continued to wait.

After a half day of meetings with local and regional officials, the undersecretary and his staff headed to the Marine Science Institute, where several hundred people again had gathered. About 40 or 50 of them were allowed into a large room; the rest waited outside. A tall man, Tony towered over almost everyone in the room. Although he exuded certainty, he was also restrained. Speaking carefully, with respect, because he knew that people were afraid, he explained what they might expect over the next few weeks. By early August, the DENR committee would decide about issuing a permit for the cement plant. "Social acceptability is not a question of numbers, of who wants what," he said. It is a process by which the DENR determines whether all the issues that have been raised—pollution, health risks, land use—have been resolved. "It is a purely technical question, a scientific question."

The idea of a decision unmarred by political fracas seemed wise. When Tony La Viña discussed this in his office in Manila and in the air-conditioned van, it had even begun to sound possible. In Bolinao, where it would become reality, it sounded precarious. People were used to one system—government officials who either ignored them or could be bought—and whether this system favored them or not, it was familiar. In this highly emotional setting, the possibility of a decision based on technical criteria seemed unreliable. Could such criteria be broad enough and flexible enough to accommodate local needs?

People stood to ask questions: What is social acceptability? Why don't you call it environmental acceptability? Who should make the decision? As they spoke up, addressing the undersecretary, they began formally and slowly in English, but as their emotions rose, they sometimes would burst out in the more personal words of their own dialect. Some tried to stay composed, some could not. The voices in the room grew slightly louder, and the several languages being spoken blended into one sound. Emotions rose more and more to the surface, and then, as a woman jumped to her feet, they broke through, spilling out like the words that tumbled into informality. Her voice was full of distress. She spoke about the plant and Bolinao and the life there. The undersecretary turned to face her across the room. She kept talking, raising first her chin, then her voice, then her hands. "Why are we discussing all this in a scientific way when the people of Bolinao don't want it?" she said. "*Ayaw ko!*—we don't want it! *That* is social unacceptability." There were more voices as others agreed. "We don't want another Marinduque in Bolinao," someone cried out. More voices rose, and the meeting ended with confusion. People had been heard but not reassured. No one knew what the decision would be.

• • •

A few weeks after the visit to Bolinao, the DENR released a statement from Secretary Victor Ramos, announcing his final decision. "Please be informed," read the letter addressed to the general manager of the Pangasinan Cement Corporation, "that we have resolved to deny with finality the issuance of an environmental compliance certificate (ECC) for your proposed Pangasinan Cement Complex project in Bolinao, Pangasinan." The committee had found the project to "pose adverse impacts to the environment which are considered irreversible and non-negotiable," particularly the environmental risks to the aquatic life and coral reefs and risks to residents' health. There were still conflicts over land and other resources. The letter also addressed in detail the "problems of social acceptability."

"The project has deeply divided the Bolinao community and the larger society of stakeholders. These deep divisions are rooted in the fundamental conflicts of interests rather than . . . mere ignorance or lack of information on the project." The letter recognized the political support the project had received as well as the "unwavering opposition." "[W]hat is regarded with equal, if not greater, importance are the issues raised against the project that are found to have remained unresolved which make the project socially unacceptable."[14]

It was clear to all involved that the Bolinao case had been long and laborious. The corporation spent several years planning the project and then several more waiting for the final decision. The town and MSI spent years organizing and negotiating the conflicts splitting the community. Tony and his staff devoted almost as long on the bureaucratic end. "This can't be done for every decision," he said at the public meeting, his voice lowered as though he were speaking to himself. He believed that a community should be involved in managing its own natural resources, and he believed that the government could make better decisions, but there were so many practical obstacles to overcome.

◆   ◆   ◆

I have been asked whether the decision over Bolinao was the right one and whether it was an environmental success story. The two are not necessarily the same, and when a local economy and its natural environment are so closely linked, it is not clear what a right decision or an environmental success story would mean. Although fishing could continue in the short term, condemning Bolinao fishermen to a life of declining catches would hardly constitute success. On the other hand, industrialization that forced fishermen to migrate again would not mean progress either. Should the officials instead have hedged their bets, as along Malalag Bay, supporting the coastal management while trying to attract new industries and outside investors? In the long term, if the Bolinao decision scared off foreign firms, a drop in investment in the country would not mean success either. A declining economy, or a stagnant one, could further hasten environmental decline.

"We didn't say no because we didn't want the industry," Tony said, reflecting on the decision recently, "but because the costs would have been too high. Some people say that sustainable development is always a matter of compromise. But it also requires saying no sometimes. On the other hand, it wouldn't be sustainable development if all of your decisions [about potential investments] were no. That would be the decision that development was not important."

A single decision, the Bolinao case could not address every question or complication. Its lasting importance was that it helped define and improve the process by which complex decisions over development were made. In a country where the government can be weak and outside investment does not necessarily benefit a community, the case set a new precedent, broadening the meaning of social acceptability to include the community's concerns.

In its attempt to invest in the country, Pangasinan Cement Corporation had cooperated and talked with the community leaders, trying to address the question of social acceptability. A high-level Taiwanese manager expressed his anger at the requirements and delays, the very local involvement that Tony and others had worked hard to achieve. He did not fully understand what social acceptability meant, and it rankled him to meet with local officials, to waste his time, to have to engage with the community. I shouldn't have to deal with those people, he said to me in an interview. "It's the government's job. The government should evaluate, decide if it's good or not—and the government should explain to the local people." While rumors persisted for several years that the corporation would reapply for a license, they remained only rumors. After such decisions, Tony said, "the company and politicians often come back and pressure us to change our minds. With Bolinao, the company said that we had been fair and that they would accept the decision."

Late in 1996, the DENR strengthened the system through which firms applied for certificates of environmental clearance, making public participation and social acceptability necessary prerequisites. Over the next few years, the DENR approved about 50 other cement and quarrying operations, including five in Pangasinan. One of them was a plant in nearby Agno on the South China Sea, but local opposition stalled the construction.

Deeper change within the agency, however, did not come quickly. The 1998 election of Joseph Estrada to the presidency brought a resurgence in corruption in the government as well as economic decline. The next secretary appointed to the DENR did not believe that communities could manage their own resources. Tony La Viña moved his family to Washington, D.C., and joined a progressive environmental think tank. Other committed officials left their government positions, and some stayed, waiting for a new administration. Estrada resigned in 2001, and under the presidency of Gloria Macapagal-Arroyo, environmental work again moved forward. Although the Philippines resisted becoming the region's sink for dirty industry, the requirement of social acceptability came to receive less emphasis.

According to one of the MSI scientists in Bolinao, livelihood continues to be a struggle, and the town has stayed much as it was, which was what the original dilemma was partly about. While Tony met with the town's leaders during our visit, I went with several of the DENR committee's members to visit an inland *barangay* where limestone quarries would be dug. One member, Emilyn Espiritu, was an environmental scientist from Ateneo de Manila University who had recently completed her doctoral studies in Europe. She was outspoken about the decision that loomed. "I've had many sleepless nights," she said. "If the plant is built, the landscape will change, the people will change, everything will change. There will be a complete change from a rural fishing village to a technological place. There's a resource waiting to be tapped. Are you going to exploit it, or are you just going to let it sit there? And if you do exploit it," she asked, "is it going to be beneficial to the entire country?"

Barefoot children ran around as we stood in the heat in our pressed blouses and leather shoes. Emilyn questioned several women about their lives and their thoughts about the cement plant. Comfortable with her, people spoke openly. In other towns, they said, factories had brought only temporary jobs or jobs too skilled for the locals, or they even restricted access to jobs. Some factories had rules that a single member of a family could work there; others hired townspeople as "casual" labor with no job security; still others would hire people for only a few months at a time. I had heard the same story along Malalag Bay and near industrial parks outside Manila. Just as the government did not necessarily benefit the people, neither did the private sector.

"Wouldn't you rather have some of the benefits from the plant?" Emilyn asked. "I wouldn't benefit," said one woman. "I wouldn't even qualify as a worker in the plant." She glanced at our clothes and carrying bags. "We're not like you city dwellers. We have simple needs. If you allow the plant to be set up, you're imposing on us your idea of what's important in life and how we should live."

7 | Why Do You Want to Know: *Bakit?*

Palawan stands alone at the western edge of the Philippine archipelago, its 267-mile mainland angling toward Luzon between the Sulu and South China Seas. Close to 2,000 islands surround Palawan's main island, many so small that they disappear at high tide. The mainland is long and straight, and the Spaniards, reportedly seeing in its spindly form a folded umbrella, called it Paragua. Later, the Americans renamed the province Palawan after one of the province's indigenous tribes. The word is thought to come from the Chinese name for the island used as early as the ninth century, *Pa-lao-yu*—land of beautiful, safe harbor.

A steep mountain chain bisects the mainland, separating the western foothills from the eastern swamplands, valleys, and the province's single sizable city, the capital of Puerto Princesa. Distant and inaccessible, until recently Palawan was seen as a humid, jungle-covered outpost where malaria was endemic and civilization did not have a foothold. Like Mindanao, the province was never successfully colonized. Having failed to subdue the indigenous tribes of Batak, Pala'wan, and Tagbanua, the Spaniards did not set up their infamous plantations that in Negros Occidental and elsewhere created a landed elite. When the Americans took control in the early 1900s, they renamed the island Palawan and built a low-security prison and a leper colony near Puerto Princesa, deepening the province's reputation as a place of exile.

Because of its geological history, Palawan has an unusual flora and fauna. It once formed a prehistoric bridge to Malaysia across which animals, fruits, and seeds traveled, creating a diversity of life that is far closer in its evolutionary origins to Borneo than to the rest of the Philippines. Biologists have counted more than 200 species of birds, more than 600 species of butterflies, and a comparable diversity of flowers. Even for the Philippines, Palawan has a large proportion of species that are unique to the province. There are more than 230 known endemic species, with such appropriate names as the Palawan bear cat, the Palawan tree shrew, the Palawan clawless otter, the Palawan stink badger, the Palawan flying fox, the Palawan peacock, the Palawan talking myna, and the Philippine cockatoo.[1]

Long after other provinces were transformed by deforestation and overfishing, Palawan remained relatively untouched. In an overcrowded country, the province was a neglected outpost with just a few hundred thousand people. Other islands had bare, eroded foothills and failing fisheries. Towering trees still covered Palawan's hills, its reefs and fisheries thrived, its valleys were lush, and rivers and streams teemed with freshwater life. When the environmental movement began to gather force in the 1980s, Palawan was believed to be an ecological frontier. It became a symbol of what the rest of the country had lost. The very lack of development also promised the possibility of economic reward.

In the early 1990s, as the Ramos administration launched plans for development that could be "clean and green," it identified tourism as one of the flagships that could carry the country toward economic recovery. The perceived benefits were numerous. Tourism could coexist more easily with fishing and agriculture than heavy industry. Where coastal fisheries and soil fertility were declining, it could bring local jobs and help draw funding for roads and other public works projects. At the national level, the tourist industry was seen as a way to link local economies with regional and global ones and attract needed foreign exchange.[2] As with other industries, the strategy was to foster regional development by bolstering the services and infrastructure in targeted tourist zones, which each had at least one priority "destination." A main zone in Luzon would be northern Palawan, its priority site the island town of El Nido, a breathtaking cluster of reef-ringed, limestone islands.

The attention quickly brought an intrusion of outsiders—not just tourists, but also representatives from nongovernmental organizations (NGOs), corporations, foreign funders, and government agencies. Their presence meant that more change was coming. It also heightened tensions over how the natural resources would be used, who would benefit from their use, how protected areas could be designed to reconcile the needs of conservation and development, and what role high-end tourism should play in Palawan's future.

•   •   •

I visited Palawan a number of times during the 1990s as the province was changing visibly, but when I first visited in 1991, the push for tourism had not yet begun. American friends and I took an 18-hour ferry from Manila headed for a pair of protected areas, St. Paul Subterranean River on the mainland's western coast and El Nido. The boat was old and overcrowded, its deck and hallways lined with beds and baggage. Like Davao, Puerto Princesa still felt like a place for exiles. Covering the middle third of the mainland, coast to coast, Puerto, as it is called, was mostly rural, without adequate roads or other services. Its center was small, with dirt roads and footpaths, wooden houses shaded by mango and banana trees, flourishing gardens, and bougainvillea tumbling over roadside fences. Modest establishments labeled with artfully hand-lettered signs lined the main road, Rizal Avenue. Motorized tricycles zipped up and down, carrying long sidecars built to hold three or four people instead of the usual two or three. Except for occasional cars and the jeepneys serving outlying areas, there were few vehicles. Life was slow and predictable, the streets were hot and dusty, and the air, while humid, was clean enough that it was comfortable to walk on Puerto's streets, as it once had been possible to walk in Manila. There were only a few restaurants and no traffic signals, and street lights glowed just a short distance down Rizal from the main commercial strip before ending abruptly in blackness. At the edge of town stood a camp for Vietnamese refugees who had crossed the South China Sea on boats, a stark, fenced-in settlement squeezed full of tiny homes.

Young spear fisherman. Busuanga, Palawan.
MARISSA ROTH

In Manila, the staff of the environmental organization that organized our trip had enthusiastically described Palawan as wilderness, a glimpse of the Philippines' lost ecosystems and lost beauty. Outside the center of Puerto, however, Palawan no longer appeared a pristine frontier. While far more intact than those in more settled provinces, the forests, reefs, and fisheries were being chiseled away. Although the vision of an ecological frontier persisted elsewhere, in Palawan itself, people knew that the rich natural life had become an arena of struggle.

Commercial interests were increasingly looking to Palawan. After the loggers had exhausted the most accessible forests in the Visayas and Luzon, they turned to more remote areas, including

Mindanao and then Palawan. Licensed logging concessions soon covered more than half of Palawan's forestland. In 1968, trees spread over 92 percent of the province; by 1989, they covered only about 48 percent.[3] Huge tracts of land had been claimed for mining, and investors were eyeing Palawan's sizable reserves of gas and oil, particularly along the western coast. Palawan has about 178 fishing grounds and nearly half of the country's coral reefs, and they faced similar pressures. With other fisheries overfished, boats from other regions were plying Palawan's coasts.

Like other places, Palawan also was attracting migrants. Their desperation showed markedly not far from the center of Puerto in a village along Honda Bay, a jetty formed of mine tailings whose land—dry, yellow-orange, and deeply fissured—was so contaminated with mercury that few plants grew anywhere. Nonetheless, the barren spit of land had become an established village covered with houses.

Palawan's population was the fastest growing in the country, doubling to more than 600,000 between 1980 and 1995. In a 1996 study more than half of the people surveyed were from other regions. They spoke 64 "local mother tongue groups," of which just nine were indigenous to Palawan.[4] Sometimes we call Palawan the "Little Philippines," a community organizer told me.

The political debate over Palawan's future had begun in the 1970s, even before the end of martial law brought an opening for environmentalism. From the start, the discussion focused on how to avoid the ecological fate met by other provinces. A preliminary study concluded, "Because the province is a unique ecological unit in the world, and the only one presently intact in the Philippines, it is imperative that conservation . . . be a major part of development."[5] A plan to guide development in mainland Palawan was completed in 1985 and then revised the following year after Marcos's ouster to include the province's smaller islands and more democratic goals.[6] Proposed to Congress a few years later, the Strategic Environmental Plan for Palawan established a new decision-making body for the province, the Palawan Council for Sustainable Development (PCSD), composed of representatives from the government and NGOs. The council was to create policies and approve development projects, decisions usually made by the DENR—thereby setting the stage for a protracted turf war between the DENR and the PCSD over Palawan's environment.

In the late 1980s, as Congress debated this plan, activists launched a campaign in Manila to pressure the national government to "Save Palawan." A signature drive entitled *Boto sa Inangbayan*, Vote for the Motherland, called for a total commercial logging ban in the province. Drawn by the "unsettling beauty," wrote journalist Yasmin Arquiza, a small group of environmentalists moved to Palawan "to help protect and preserve this piece of paradise on Earth."[7] In time they would form dozens of NGOs—some focused solely on ecological matters, others intent on also working with the indigenous groups—and become part of policy making and governance.

One of the first environmental NGOs to be set up in Puerto was Haribon Palawan, an offshoot of the national Haribon Foundation for the Conservation of Natural Resources, the country's oldest environmental organization. Haribon was created as a bird-watching group in 1972, its name derived from the Tagalog words *hari* (king) and *ibon* (bird), roughly translated into English as King of Birds, the endangered Philippine eagle. Its founder and former DENR undersecretary, the physicist Celso Roque, believed that the country needed a new environmentalism that would do far more than simply mirror its western counterpart. "An appropriate environmentalism," he wrote, would be tailored to the "social and ecological landscape" of the Philippines, which is tropical and largely agricultural, marked by poverty and a growing population, whose access to natural resources is "feudal rather than democratic."[8] Under his leadership, Haribon evolved into a conservation organization that worked with local communities.

The Palawan branch of Haribon was started by anti-Marcos activists who came to the frontier from Manila. The organization's head, Joselito (Lito) Alisuag, was an outspoken young attorney who, like other founding members, had previously worked with the labor and human rights movements. By 1986, said Lito, people in Palawan were "already concerned about the environment, concerned about the logging and floods." When the "Save Palawan" campaign started, Lito, then a recent law school graduate, volunteered to help organize Haribon. Speaker of the House Ramon Mitra, the province's powerful congressman, "was the kingpin then, thought to be the next president," said Lito. "He was close to the loggers and to the church also. No one was willing to lead because the adversary was so powerful. It would take someone who was crazy, so I had to do it."

Beginning with about a dozen members, Haribon Palawan first focused on illegal cutting, which was flourishing in Palawan. Almost all the politicians here "are involved with illegal activities" such as logging and fishing, said a senior DENR employee. "If you're doing a good job here in Palawan, if you're succeeding at environmental protection, they brand you as a communist." Using their political skills, Lito and others went to organize rural villages, where they met less suspicion than organizers had on Apo Island. "People listened to us," Lito said. "We had been activists against Marcos, so people believed us." In time, though, those credentials worked against them: "Our background caught up with us. We were noticed, and the harassment started. Mitra needed to silence his province."

• • •

Tensions reached a new pitch in 1991, shortly before my first visit to Palawan. My friends and I expected to be hosted by Haribon Palawan, but just before we arrived, 14 members were arrested, and others went into hiding. Lito, who had been in Thailand, was arrested at the airport when he returned. Several Haribon members, shaken, explained what had happened. Early in January, three of them were working in southern Palawan on a development project with the Batak, a forest-foraging tribe, when they chanced upon a huge cache of *kamagong*, Philippine ebony. A tree with hard, heavy wood, *kamagong* is called "black gold" because of its deep color and the price it brings. Although legally protected from logging, *kamagong* was being smuggled in speedboats to Malaysia, where it was in high demand. Loggers typically cut freshly felled logs into 10-foot lengths called flitches, which are easier to ship. One tree might yield three or four flitches. The cache contained 300 to 400 flitches, worth about 120,000 dollars. It was about 100 yards from a military detachment.

According to a staff member who had been arrested, "One of the community organizers went to conduct a social investigation and answered the call of nature. He saw the logs covered with leaves. He took pictures, inspected them. They were choice cuts." The organizer reported the illegal wood to the military, which failed to respond. Haribon then "exposed it" to the DENR office in Manila. In mid-February, the arrests began. According to others, "heavily armed men in full gear" came to their homes at dawn and took them to the military camp. Before they were released, they were asked to return the next day; when they did, they were detained a second day. Three of the fourteen were Haribon officers, and most were in their 20s and 30s, married with children. Lito was named as the "so-called chair of the Palawan Communist Party" and charged with subversion and rebellion.

The Batak also faced intimidation. According to the *Philippine Daily Inquirer*, after the *kamagong* was discovered, "unidentified Marines" harassed tribesmen in the village and "confiscated their *bolos* [machetes] and farm implements." Some 300 Batak families threatened to commit suicide if the harassment did not stop.[9] A Haribon organizer explained: "In their culture, it's wrong to fight back. It's more honorable to kill oneself than to engage in conflict. Their houses had been ransacked, some of the men were beaten up or hung for several minutes with their toes touching the ground. They walked 186 miles, going across the mountains to avoid the highway and military, to ask us for help."

The crisis was short lived. The tribal families did not commit suicide, and all of the charges eventually were dropped. Later that year, the flood in Ormoc further awakened people to the dangers of overlogging. Reflected Lito, "People realized that what happened at Ormoc could happen here."

In 1992, Congress passed the Strategic Environmental Plan, which Corazon Aquino signed as one of her last acts as president. That year the green vote also helped elect a handful of new politicians, often called born-again environmentalists, including the governor, Salvador Socrates, and the mayor of Puerto Princesa, Edward Hagedorn, who ironically came from a logging family. Together with the Strategic Environmental Plan and the newly formed PCSD, these politicians moved the struggles over the environment and development to within the government. In time, Haribon Palawan became an ally of Hagedorn, helping him launch environmental protection programs to stem illegal logging and fishing in Puerto. Within the decade, Lito ascended to head of the technical staff of the PCSD, the research arm of the body charged with implementing the province's environmental plan.

After the arrests in 1991, however, no one could have predicted this future. The Haribon staff, agitated and afraid of being seen with foreign guests, would speak to us only indoors. Before we left, they gave us a letter of introduction to take to Nilda Baling, who worked for the DENR and at that time still served as the influential head of the marine reserve in El Nido. Nilda represented one path—toward protected areas that would stress conservation over tourism and other economic development. With Marcos's cronies still controlling the rural provinces, battles over conservation were fought mainly at the local level. Over the next few years, as the new generation of politicians was elected, ecological protection would become part of public debate. Tourism to Palawan would also more than quadruple, and the local struggles over the province's protected areas would swell into full-fledged, high-stakes conflicts over this lucrative industry, the roles that the government and private sector would play in its development, and the question of which environmental laws would have jurisdiction over Palawan.

⚘

El Nido lies along Bacuit Bay on the western edge of Palawan, just before the northern tip gives way to a spray of islands. It lends itself to hyperbole, a tropical paradise with white sands, luminous reefs, and haystack-shaped islands scattered across the calm, blue-green water. The town proper covers a narrow strip of coastal land, a border between the bay and an imposing limestone wall. Beyond the cliff, the *barangays* extend into the forested inland, and the town also includes the tall islands dotting the bay. El Nido is memorable. It is a place that is so humid that wet footprints left crossing a porch's painted wood floor in early evening will not dry by morning. It is a town without a bank, a place at the edge of settled life.

In 1991, my friends and I traveled to El Nido the slow way, taking jeepneys from Puerto up to Taytay, the nearest large town. The road had recently been improved, but within a few years, the tropical rains would again tear up its surface, and what took us much of one day would take a day and a night or more. Our jeepney was the only one to head north that day, and it was overloaded. We perched on the roof for the last leg of the journey, ducking low-lying vines and branches and marveling at the forested landscape.

The logging concessions were operating in full force. The road crossed eight to ten streams, and at each one, the small bridges over the water had been raised several feet and reinforced with wood. When I inquired why, I was told that they had been built up to bear heavy logging trucks, particularly during the rainy season: an election was coming up, and logging would help the politicians fill their campaign chests.

The commercial loggers had felled only the larger trees, but their roads opened up the forest. As the jeepney followed the road through the forest, we passed evidence of migrants. Freshly cleared, blackened plots abutted the road all the way north, and telltale curls of smoke from burning brush spiraled upward above the trees. On the return trip, a half-day pumpboat ride from El Nido to Taytay, there was never a time when, in some direction, I failed to see a fire burning on an island or a plume of smoke curling above the horizon where land was being cleared.

One of the oldest towns in Palawan, El Nido used to be called Bakit, from the Tagalog word for *why*. When the Spaniards came, the story goes, and asked the locals questions, the reply they got was *Bakit?*—Why do you want to know? The town was later renamed El Nido, which means "the nest" in Spanish. Its many limestone islands, called tower-karsts by geologists, have steep walls that have weathered into caverns and hollows. In these cavities live colonies of *balinsasayaw*, edible-nest swiftlets, sleek birds that soar across the sky in dramatic arcs. The swifts use their saliva to build and cement half-cup nests into the limestone hollows. Prized as delicacies that can convey health, energy, and sexual vitality, the white nests are sold as a medicinal remedy and as the key ingredient in bird's nest soup.

In the past, traditional gatherers, *boceadores*, collected the swiftlets' nests, passing on to the next generation the precarious skill of scaling the cliffs and cave walls. To ensure a continuing supply, they did not gather nests bearing eggs and kept a vigil, often in a house on site, to safeguard them until the nestlings hatched. The nests used to be the town's main source of income; a first-class specimen might bring 1,000 dollars. In the 1990s, as demand for the nests rose, largely from Japan, poachers came, joining the other outsiders hoping to profit from El Nido.

In 1991, in this town of 8,000, there were still few outsiders and just a single resort, Ten Knots, on an island in the bay. While the resort had brought a few jobs, life on the mainland remained relatively unchanged. Early in the morning, at one end of the shore, *bancas* were pulled up on the sand and tiny fish splayed like stiff butterflies lay drying on wooden racks. At the other end stood a string of visitors' cottages, modest dwellings often built between the water and the even simpler home of the owner's family. The town's generator had broken, and there was no electricity. There would be none for two more years, and even then, the generator would only run from six until midnight and send power to few of the outlying *barangays*.

El Nido had one main road, a broad sandy lane where young schoolgirls wearing white blouses and long, pleated maroon skirts walked before breakfast carrying strings of fish. Two girls carried theirs on a pole balanced across their shoulders. The market was small, and on an ordinary day, by afternoon only a few stalls were open, offering pale yellow finger bananas, squat green cooking bananas, a few meter-long *sitaw* beans, some gold *calabasa* squash. Other than this limited produce and the day's catch, everything in El Nido, from vegetables to eggs, had been brought there by ferry, plane, or jeepney.

As the fortunes of El Nido changed over the next few years, a covered market was built, and stores filled with stacks of shorts and shirts and cropped pants of bright Indonesian fabric. Water remained limited, as did electricity, but there were more island resorts, cottages, and restaurants to accommodate the growing numbers of tourists. By that time, people no longer caught and split fish to dry because they could earn better money using their pumpboats to transport tourists. By then, that edge of the South China Sea was no longer being watched over by the determined biologist from the DENR, Nilda Baling.

. . .

Nilda had worked with the marine reserve in El Nido since 1984, when it was created by the World Wildlife Fund as part of a large debt-for-nature swap, a scheme to help developing countries finance the conservation of tropical forests while reducing their foreign debt. The concern was that countries were relying on exports of logs to repay their loans. By buying part of an outstanding debt and using the funds to set up conservation projects, the World Wildlife Fund hoped to deter deforestation. From 1989 to 1992, in the first such swap in the Asia-Pacific region, the organization bought up to 2 billion dollars in Philippine debt, providing several years' funding for protected areas at El Nido and St. Paul Subterranean River, where the Cabayugan River flows five miles underground through dripstones, stalactites, and frozen waterfalls.

A 17-year veteran of the DENR, Nilda had played a major role in designing St. Paul, a park she called "my baby." To her surprise, rather than being assigned there, in 1989 she was transferred to El Nido to establish the sanctuary in Bacuit Bay. "What?" she recalled saying. "I don't even know how to swim. At first I didn't see it as an honor. I'm not from a coastal area. I was a stranger to the place."

Nilda and her husband moved to El Nido, expecting to set up the office and stay for three months. She ended up running the project for seven years and managing 34 people. While many urban government employees dread being transferred to the provinces, some love the quieter life there. Nilda was among them, and she soon came to see El Nido as their home. She also took seriously, far more seriously than ended up being expedient, the challenge of protecting the reserve. Because of her relentless pursuit of illegal fishermen and loggers, before long she became known, both behind her back and to her face, as the Dragon Lady.

For a dragon lady, Nilda appeared mild. Polite and soft-spoken, she punctuated her speech with small, short laughs, often pursing her lips delicately, with a downward glance. Her quietness masked toughness and decisiveness, and while the staff were comfortable but respectful with her, they moved quickly and almost in unison to respond to her requests.

Surrounded by the calm, aqua waters of Bacuit Bay, El Nido seems a restful place. The islands swell up from the seafloor, their steep sides matching grays. Some islands are little more than limestone chunks; others have a forested dome and border of flat shore. Some are fringed by a bit of sandy beach; on others, the persistent lapping of the surf has carved away the limestone, forming caves where, at low tide, swimmers can duck beneath the island's overhanging lip and rest in its shadow.

We spent a day motoring from one to another of El Nido's islands and diving on the reefs. Several of the DENR staff joined us, as did Nilda's husband. After our lunch of rice and grilled fish, Nilda described how they created the marine reserve in an area rife with illegal fishing. She often answered questions with the plural pronoun, which included either her husband or her staff or both. "Before we came," she said, "enforcement was lax." They started by protecting the coasts and reef fisheries; they prohibited the collection of shells, corals, and tropical fish and cracked down on the dynamite and cyanide fishing.

The marine reserve, which at first covered 19 islands, resembled other protected areas. The town itself and islands closest to it made up the core zone, where few activities were permitted. The surrounding islands were part of the buffer zone, an area for mixed use: some fishing was allowed, but trees could be cut only with permission, and sand and rock could not be mined. An area that large defies monitoring, and patrolling the waters and apprehending fishermen were dangerous as well. The rangers always carried guns, and, a restaurant owner told me later, they would pack their boat with sandbags to protect themselves from gunfire. Some environmentalists then were motivated by political conviction, and others were sustained by religious belief, including Nilda. She recalled being threatened "many times. I've been offered 30,000 pesos [about 1,000 dollars] and a carbine to stop our campaign against illegal fishing. Illegal fishermen even follow me to Manila. I'm brave because of my commitment to God. He guides me, helps me. As long as I'm doing the right thing, I wouldn't be afraid."

As we ferried among the islands, Nilda recounted their years there, calmly pointing to specific spots to place events—a stretch of water where dynamite fishers had fired at them; another stretch where they caught the men; a long forest on the mainland that burned before they arrived. On one island, in half-cleared plots, the staff had apprehended would-be farmers. As we passed another island, a man, bent over hoeing in a new clearing among the trees, looked up with what seemed resignation. They would look for him later, she said.

Although the reserve was established to safeguard the marine area, Nilda knew that the coasts could not thrive without forests, and she was fierce—dragonlike—in her pursuit of both illegal fishermen and loggers. A huge logging concession still covered El Nido's forests. The loggers claimed to abide by laws prohibiting them from cutting such trees as *kamagong* and any trees below the legal limit of 2 meters or 25 years old. We saw otherwise. At a log pond, a stack of round logs waited to be loaded on boats. We cautiously pulled up. The pile included protected species as well as many logs narrower than 1 meter across that were probably less than 10 years old. With the Strategic Environmental Plan and its logging ban close to becoming law, the loggers were cutting fast, "speeding up," said Nilda, "to get the undersized trees."

While she and her staff had controlled illegal fishing, they were less successful at stopping the illegal loggers and *kaingiñeros*, who blamed one another for the forests' destruction. She visualized the logging roads stretching ever deeper into the forests, like tentacles: "Wherever the trees are, that's where they develop the roads. It's an octopus. The forest would return, but then the *kaingiñeros* take over. They burn everything. We can't control them either. We have to remove the logging, they say, 'before you touch us.' That's one of my failures here—the logging. We were able to control it but not stop it."

El Nido's forests were disappearing at a pace that was rapid even for Palawan. In 1985, nearly half of El Nido reportedly was covered by old-growth forest, which by 1992 had dropped to less than 10 percent. Studies of Bacuit Bay showed that loss of the forests ultimately harmed the reefs and coastal fisheries. The logging roads were constructed carelessly on steep slopes that eroded quickly. During strong rains, runoff from the roads flowed into the coastal waters, carrying a heavy load of soil.[10]

"The effect of sedimentation is like cyanide," said Nilda. "It's only now—after two years—that the corals are sprouting and regenerating." She was referring to coral's vulnerability to sediment. When silt drifts onto a reef, individual corals clean themselves by moving as whole polyps or by wiggling their cilia—a miniature, underwater version of cows swatting flies with their tails. Too much silt effectively smothers the corals. In 1996, biologists Gregor Hodgson and John Dixon estimated that during an eight-month period, 128,000 tons of sediment washed into the bay, much of it from the logging roads and mainly during one strong storm. In the reef closest to the mouth of the river that drained the logging area, they found that, because of the silt, nearly half of the corals had died.[11]

Nilda did not prepare us for the biological glory that lay beneath the calm sea. Palawan reefs are among the country's most spectacular, and great mounds of finely architectured coral covered the ocean floor. Life bloomed in colors ranging from gold to orange to electric blue. The corals' delicate tentacles undulated and shivered, and schools of fish angled about in orchestrated movements. Nilda was proud of the healthy reef but, wanting to show us both the best and the worst of El Nido, had the boat drop anchor in an area that had been hit by dynamite. The stop was brief. In the half-light, as far as I could see in all directions, the ocean floor was pale with shattered corals, another stretch of reef lost.

"The dynamite fishing is done not by local people," said Nilda, "but people from the north and south [provinces], mostly Batangas and Cavite. Some fishermen are financed by big military men. Our rangers are armed, and most of the fishermen are armed. One of our rangers was shot." More boats were coming to Palawan's rich fishing grounds, many of them large, commercial boats whose hauls ended up in Manila, where they provided more than half the fish resting on ice in the city's wet markets. Preventing the illegal fishing proved to be a challenge, although the local fishermen were starting to understand the value of the reserve. Without the protection of the military, Nilda and her staff had to seek the aid of the local militia, the Civilian Armed Forces Geographical Units (CAFGU). "It's hard to get the support of the military, who receive a low income and are easily corruptible," she said. "We pay 900 pesos [about 35 dollars] to four CAFGU on a rotating basis to protect the waters and give a show of force to the rangers. Now people believe that we are protecting the corals so there will be more fish for them to catch, so we have their support." Still, she admitted, "if there were no project such as this reserve, it would be a free-for-all again."

Nilda also faced opposition from politicians and businesses in El Nido, and her efforts to prevent illegal logging ultimately undid her. Fishing was the main source of livelihood, and *barangay* officials supported her, but those at the municipal level did not. They wanted the town to develop, and they saw Nilda as a key obstacle; each new hotel or restaurant required wood whose cutting she hindered. She opposed as well the construction of a better road, which would dump huge quantities of silt along the coast.

Without more political support, Nilda could offer limited protection at best. "We report violations, like the construction of logging roads and cutting of undersized trees, but [when there is an investigation] the findings favor the loggers," she said. "We've sent about 50 citations to the DENR, maybe half through the municipal office, and none of those submitted through the municipal government were actually passed on and reported. Sometimes the municipal government reports that we're fabricating cases. When we apprehend someone, they try to mediate for their release. Sometimes we are like fools, arresting while they free them, so the violators laugh at us. The intervention of politics is killing our work."

. . .

El Nido is missing from maps of the Philippines drawn as recently as the 1970s. According to local legend, a dive boat discovered the town accidentally in 1979 when it snagged a fishing line in the middle of the night and had to drop anchor. The next morning, divers awoke to the arresting island landscapes, and tourism began. Ten Knots was the first resort, built in 1982 on the island of Miniloc, about a 30-minute boat ride from the town proper. Set up as a joint venture between a Japanese sugar manufacturer and the Andres Soriano Corporation, the resort also marked the entry of the powerful Soriano family into tourism in northern Palawan. The Sorianos, one of the manufacturing families whose wealth came from the sugar industry, produce soft drinks and glass bottles. For decades they ran the massive San Miguel Corporation, whose pale lager is a mainstay of drinking life in the Philippines. By the mid-1990s, they would also own three of the four resorts inside the marine reserve.

Nilda had a vision of fostering tourism in El Nido that would be small and homegrown. Rather than building more cottages, people could "improve" their own houses to open them to tourists. Ten Knots was a far cry from this vision. At the edge of the cove, a row of thatched cottages stood on stilts, reflected in the still water. Luxurious and orderly, the resort first attracted mainly Japanese tourists. In a town with persistent water shortages, where most buildings lacked plumbing, the bathrooms had running water and chrome fixtures that shone.

In time, the relationship between the Sorianos and the DENR deteriorated. "The resorts violate laws," claimed a DENR employee. When the reserve was created, Ten Knots was grandfathered in, but the Sorianos, he said, built another resort before receiving the needed environmental clearance from the DENR. "They paid local people for the lumber," he said, and "quarried white sand from other islands, another violation." A few months earlier, Nilda and her staff had apprehended the owner of the land where the lumber had been cut illegally. "The Japanese resort depends on him for their lumber needs," Nilda said matter-of-factly. "We can only apprehend the possessor, not the perpetrator. We seized the lumber, and both of them lost money. Ten Knots hadn't paid for it all yet. I was threatened afterwards, was advised by friends not to come back. A killer had been hired."

Over the next few years, Nilda steered the sanctuary through several major transitions. After the funding from the debt-for-nature swap ended, the reserve funds she had carefully built up carried the project for an additional year. When new funding ran out in 1995, the staff worked without pay for about six months. Some then took part-time jobs but continued to volunteer. Nilda had trained local people to work in the marine reserve, believing that in time "the park would be turned over to them." She did not appreciate fully the economic and political power of Ten Knots.

In 1996, she was abruptly recalled to the central DENR office in Quezon City. Without funding or her leadership, the project unraveled. The island research station, a handsome two-story wooden structure, was stripped of nearly every removable part, from windows to floor planks to plumbing. As she had feared, in the reserve at large logging and fishing quickly resumed. A city council member said bluntly, "It's a fiesta now."

In Manila on a research trip, I telephoned the DENR looking for a way to reach Nilda and was surprised to be told that she was there. The agency is one of a series of blocky government buildings that line the boulevards angling off the large, grassy roundabout of the Elliptical Circle. Deep in the labyrinth of the immense complex, surrounded by open offices filled with mazes of desks and government workers wearing matching pastel uniforms, I found Nilda. She was in involuntary exile, temporarily relieved of her duties in El Nido by then undersecretary Delphin Ganapin. She seemed out of place, if not ill at ease, indoors.

"The government and business—the resort owners—are pressuring the DENR to remove me," she told me quietly. "My effectiveness is becoming a liability. I was questioning the building of roads and bridges; they're sourcing lumber there for bridges and roads, and the gravel comes from the river-beds. They said I was preventing the development of Palawan. They want to develop—there's pressure from the private sector."

A DENR official gave two different versions of why Nilda had been recalled from El Nido. On the record, he said that the project "had terminated"; off the record, and with obvious empathy, he said, "She stepped on so many toes. She's good, but as a manager she should have been able to relate to different stakehold-ers, should have been more flexible so she didn't raise the ire of others. She had conflicts with the DENR, with Ten Knots." The official paused. "She did lots of confiscation, filing of cases against loggers and fishers, but she didn't do IEC"—information and education campaigns, a phrase usually used for public relations drives—"and she didn't mingle with the local officials." Others spoke less deli-cately about why Nilda had been replaced. "She was recalled because people were going to kill her," said one DENR employee, a conclusion that others echoed.

On Mount Kitanglad, Apo Island, and Malalag Bay, conservation focused on small-scale livelihood. In northern Palawan, with the expectation that tour-ism would flourish, the approach to conservation was changing to include the powerful commercial stakeholders. Rather than having the government man-age the protected area, the new strategy was for the resort owners—the Soria-nos—to take on the responsibility for conservation so that revenues from tour-ism could pay for environmental protection. A Soriano employee echoed the new philosophy. "This is the product we're selling. We have to safeguard it," he said. "There was animosity among the groups before. There was too much emphasis on enforcement."

Out of step, Nilda still believed in strict enforcement. "The Sorianos were telling the community that I was frozen by the DENR," she said. "*Hiyâ*, I was shamed. I asked what I had done. The undersecretary says the governor is mad at me, but I don't think he would be as angry as the resort owners, who see me as someone who would block their plans. He banks more on the pri-vate sector to do conservation, says that they benefit more from conservation and should be motivated to do it. He says that I should remember that Sori-ano is the major stakeholder in the area. But you can't see all the stakehold-ers as equal—whom should you focus on? I would rather focus more on the community and how they'll benefit."

She spoke sparingly about her duties in Manila, just once mentioning a "clean-up day" held in the parks. "We spent tens of thousands of pesos on T-shirts and sun visors. We did nothing." She talked far more about El Nido: how she wanted to return, how land prices were rising, how the community was not ready for a huge influx of tourists. There were too many stakeholders, "each dic-tating its own agenda. Everything is at a standstill." And there were too many outsiders coming in—from the government and foundations and NGOs—with too much foreign money. "They bring confusion and false hopes," said Nilda, "and dictate things that aren't the desire of the community."

⨎

The poet Yehuda Amichai captured for locals everywhere the experience of living in a place to which others flock for pleasure. Amichai, who spent most of his life in Israel, wrote of tourists and hikers for whom the memories and history of a landscape, the reenactment of a battle, might overshadow any lives currently unfolding. A poem set in his beloved city of Jerusalem, titled simply "Tourists," tells the story of a man who crosses paths with a group of sightseers. On his way home after shopping at the market, he stops near the gate of David's Citadel with his heavy baskets. Nearby, tourists are listening to their guide talk about a Roman arch, which rises just beyond his head. As the man rests, he realizes that he has become their "point of reference"; when he moves, he hears someone complain. "Redemption," he says to himself, "will come only when they are told, 'Do you see that arch over there from the Roman period? It doesn't matter, but near it, a little to the left and then down a bit, there's a man who has just bought fruit and vegetables for his family.'"[12]

Tourism on a mass scale is a modern phenomenon, borne of urbanization, affluence, and leisure. The social scientist Dean MacCannell, in a classic book on tourism and the leisure class, describes tourists "scaveng[ing] the earth for new experiences" in an attempt to overcome the alienation of modern life.[13] This search for diversion—which he recasts as a search for connection—has fostered one of the world's largest industries. Governments of many poorer countries have welcomed tourism as a quick path to sustainable development, a way to grow economically and still conserve their natural resources.

While tourism, particularly ecotourism, can bring development that is less ecologically destructive than industrialization, it, too, can degrade the very surroundings that make it possible. Tourists do not see the toll they can take on the natural environment or appreciate how many resources they require—electricity, water, food, sewage treatment—just to relax. Travelers and their money can divide a local community or lead people to see their own culture as inferior to wealthier ones. The economic growth brought by tourism may be uneven, enriching resorts, hotels, and restaurants while leaving the rest undeveloped. Over time, a community's reactions to tourists can evolve from gratitude to apathy to annoyance to antagonism.[14]

In the Philippines, tourism also raises fears, and in Palawan, Puerto Galera is what people fear most. A town on the island of Mindoro once known for its white beaches and diving, Puerto Galera became a favorite spot for divers and male tourists and was transformed into a kind of Filipino hell, a fishing town gone gaudy, overrun with bars, dance halls, and a constantly replenished supply of foreign men. There, white men of all ages and primped girls from the provinces—their long hair coifed, their new clothes tight and revealing—circle in a sad choreography of flirtation and pretense.

There is a century-old slippery slope in the Philippines leading from tourism to a sordid sex industry that envelopes men, women, and children. In 1902, at the start of the American era, Governor-General William Howard already reported "houses of prostitution" at every military post in the Philippines, and the U.S. military supported the presence of brothels.[15] The military presence continued, and so did the prostitution. Over the past century, Filipinas working in the sex industry have been legally dubbed "entertainers," "hospitality girls," and "comfort women," the term of choice during World War II. More recently, the women have been called "guest relation officers," although the GROs, as they are known, may not actually leave their place of employment to "go out" with men. To keep their licenses, women in the sex industry must get regular checkups for sexually transmitted diseases at a network of government clinics that date back to the American Social Hygiene Movement of the early 1900s, which attempted to blot out the causes of prostitution and other crimes. During World War I, the U.S. government set up comparable clinics overseas in an effort not to end prostitution, but rather to protect its troops from venereal diseases.[16] In the Philippines, the social hygiene clinics remained.

During the Vietnam War, the towns adjacent to Clark Air Force Base and Subic Bay Naval Base ballooned as seedy rest-and-recreation centers, drawing tens of thousands of women, usually from the poorest—and most ecologically degraded—rural provinces, to work in bars, clubs, massage parlors, and brothels. Until the U.S. military left the two huge bases in 1991 and 1992, it funded HIV testing, the AIDS hospital unit in Manila, and the towns' social hygiene clinics. I visited several of these surreal places, where women waiting in long lines to be tested for venereal diseases would be herded through the clinic in droves, stuck with needles, and "swabbed."

The mushrooming of tourism in the Philippines also had other roots. In the early 1970s, writes political scientist Linda K. Richter, the Philippine government pushed to expand the tourism industry rapidly as a way to whitewash martial law to the rest of the world. Marcos wanted to thwart opposition to his leadership, especially among Western democracies, and encourage foreign investments and aid. Under the influence of Imelda Marcos, who aspired to make Manila "blossom into an international oasis for the luxury traveler," five-

mother-in-law spoke to both of you last November about the above matter," it read. The writer "would like to finalize this"—the decision about the resort—before leaving the country in two weeks. "This is not a formal letter, and there's resentment on the part of the staff," the researcher said upon returning a few minutes later. Some "feel pushed" by the head of the PCSD and the governor.

Hardly unusual, that case and others also heightened tensions between the PCSD and the environmental agency. One controversy surrounded Matinloc, an island near the Ten Knots resort. A small group, purportedly a Christian organization, had occupied Matinloc since 1983 "in the concept of an owner," read a petition submitted to the DENR. On the spot where a Frenchman allegedly saw a vision of a church descending from heaven, they built a wharf, a center, and a gleaming marble shrine with a revolving statue of Mary. Calling itself the Movement of Mary, the group proposed building a resort as well, although Matinloc lies within the reserve's core zone, where most activities are restricted and new construction is barred. Allegedly religious, the group was made up of "high-ranking generals" and their wives, and a special military detachment guarded the island, said the local DENR employee who took me there.

The DENR denied the petition, but the governor, the mayor of El Nido, and the head of the PCSD supported it. The "owners" proceeded, sinking foundations for a pier and building, which officials in the town proper were unaware existed. "The DENR issued a cease-and-desist order and stopped the construction," said another of its employees, "but they completed a concrete structure of three stories. It's really out of this world. I don't know what the DENR will do. It's hard to stop them."

The PCSD staff researcher, who had been the team leader for this case, cited it as another example of how projects were being approved without the technical staff's go-ahead: "The staff voted no, but the PCSD issued an endorsement. It was personal politics. I got angry. Our evaluation meant nothing. I feel cynical sometimes." It was not unusual, he said, for construction of a project to begin before it had been approved by the necessary government agency. "That's how it is here," he said.

⚜

By the mid-1990s, several airlines were plying the route between Manila and Puerto Princesa with full planes. From the air, it was easy to see that the forest was dwindling, and the once-dense expanses of green appeared moth-eaten where patches of thick trees had been cleared. In Puerto, there was no traffic light, but banks had been built, and during rush hour, motorized tricycles backed up for blocks along Rizal Avenue, filling the air with fumes. Water shortages and brownouts punctuated daily life, yet the city was growing, every year adding more banks, more electric or neon signs, a new concrete supermarket. The refugee camp had closed, and along Rizal, enterprising Vietnamese had opened noodle houses that sold short crusty loaves of French bread and steaming *chao long,* noodle soup, with lemongrass and mint. Puerto already had many more tourists than just a few years earlier, and new discotheques along Rizal enticed customers with promises of GROs—female companions. The number of people involved in the "flesh trade," a local paper reported, had "soared to 300," not counting the GROs in karaoke and sing-along bars.[22]

In northern Palawan, there were more and more outsiders angling to benefit from the final frontier. Plans to develop the area, including the construction of roads and an international airport, continued to move ahead. Not far from El Nido, a huge reserve of natural gas had been found in the Malampaya Sound, and Shell Philippines Exploration, with approval from the DENR (but not the PCSD), had begun preparations to tap it.

In El Nido, the protected area had been expanded, and several donors were funding separate conservation projects, with mixed results. Guests arriving at Ten Knots received a code of environmental conduct, a list of "El Ni-*do*'s" about how to treat the coast: "*Do* admire the corals but remember that they are very delicate. *Do* help keep the waters of El Nido crystal clear. *Do* make a personal commitment to live an environmentally ethical life." The voluntary codes of conduct—and the reliance on the resorts to do conservation—had had limited effect. Biologist Gregor Hodgson, who studied the effects of logging on the corals in the 1980s, reported, "They've done a good job of preserving the reef in front of the resort, but the rest has been degraded and fished out. The big resorts have been environmentally friendly, but it wasn't enough. The corals are okay, but the fish are gone."[23]

The resorts had expanded, but the luxury and pleasures that visitors experienced there masked the realities of daily life for El Nido's residents. The owner of a large restaurant in town whose family had moved there in the 1970s revealed that most of the water used in the restaurant was hauled down from a mountain spring. "Everything's a problem here," he said, "the water, transportation, communications, electricity, and we have no hospital."

Nonetheless, the guesthouses were bursting with representatives from donor agencies, the resorts, and NGOs, who over meals and beers—and through evening brownouts—excitedly discussed El Nido's future. One European representative arrived with his son and daughter, blond, strapping children in their late teens or early 20s. They came as tourists. As children watched, the young woman sauntered along the beach in a top with a plunging back. I passed the son hanging from a pedicab, oblivious as the diminutive driver strained to balance the vehicle and peddle it forward. There's a man, the Israeli poet Amichai might have written about the driver, trying to buy rice for his family.

The presence of so many outsiders—what Nilda called too many stakehold-ers—began to stymie the local people, who eventually challenged the town's designation as a protected area. A European Union study described "a breeding ground for conflicted vested interests and duplication of effort. Worst of all, the different activities confuse local populations, who soon learn to resent interfer-ence in their daily lives."[24] Tourists continued to arrive in growing numbers. "In 1985, there were a few houses on the beach," said Hodgson. "Now there are little shacks up to the point and back to the mountains." Cottages had been built on stilts out into the water, and their raw sewage had contaminated the coast.

•   •   •

Many Palaweños took pride in the beauty of their province and the way it drew others. "We call our place, 'Come Back Come Back,'" a government forester called Emy told me. Once people come here, she said, they have to return. Emy had lived in Palawan for 20 years, more than half her life, and she considered herself a Palaweña.

"It's been very peaceful here," she said, meaning that it had been free of mili-tary conflict. She also used the word in a more direct sense to describe a quiet, simple life. It's true that life was hard in Palawan, she acknowledged. Supplies of water and electricity in Puerto were so limited that people hesitated to admit how few hours a day they had either. Emy rode a motorcycle, more affordable than a car, and she lived with her extended family in a home with no phone. Like the pastor in Bolinao, she was not sure she wanted a different life. "I'm old-fashioned," she said. "Things are changing, and they're changing too fast." The quiet life might not be what everyone would want, but it was why many people had come there. It was, she said, what they wanted.

Too many others, it seemed, also wanted a piece of Palawan, particularly El Nido—the donors, the local government, the resort owners, the tourists, the journalists. The town was being marketed to the world as an island paradise, but under pressure from those who would pump up the tourism trade beyond what the town could support or who would detonate the ocean for a good catch, the very thing that made the place idyllic—the protected area that Nilda had created—was disappearing. El Nido and the rest of Palawan gave Filipino environmentalists hope, and the idea of a pristine Palawan that reflected the country's past beauty and ecological richness was hard to relinquish. But the outsiders could not sustain what Nilda had begun, and they, too, contributed to its transformation.

One writer from Manila chronicled her disillusionment. Hoping to discover the provincial life, Criselda Yabes and her boyfriend had found land in Pasadeña, a village about eight and a half miles from El Nido's town proper. "We bought a piece of land in the village, behind the elementary school by the pastureland dotted with cashew trees. . . . It was purely paradise." They wanted to garden, they wanted to "[watch] something grow," and "there was something special in the sky over Pasadeña." They also wanted, she wrote, to help preserve El Nido, save it from foreigners, and teach villagers a bit of environmental sensibility. Yabes, who braved coup attempts against Corazon Aquino to write a history of the Philippine military, understood power and corruption in her society all too well. She still dreamed that El Nido might be unmarred.

Knowing the stories of *bayanihan*—the spirit of cooperative work that united rural communities—Yabes and her boyfriend imagined being welcomed, receiving help constructing a *nipa* hut. Instead, after being swindled by a realtor from Puerto, they found that a provincial life and the *bayanihan* spirit were not theirs to share. Speculators were rapidly buying up land in Pasadeña, and their neighbors wanted more money to help build a basic house than the couple had paid for the property. People were unfriendly, and at night, buzzing chain saws disturbed their sleep as men "massacre[d] the forest."

Yabes's disappointment could be dismissed as that of an urban environmentalist whose romantic view of rural life might help condemn rural people to poverty. She saw clearly, however, that when people need development—access to electricity and water and wages—they often grab whatever opportunities they can to escape poverty, even if doing so could ruin their future.

"In Pasadeña, nearly everyone we've come across [has] either sold a piece of their family heritage or [is] putting up their land [on] the market," wrote Yabes. The two were too late. They returned to Manila with a different story to tell, not of the *bayanihan* spirit, not of the camaraderie of rural life, but "of what has become of Palawan—the island we so proudly publicize as our Last Frontier." An ecological paradise was not necessarily a human one. She had imagined this paradise, she realized: "I should have known better."[25]

Part III

Finite Land and the Urban Frontier

8 | Nameless Solidarity: *Quedan Kaisahan*

The northern end of Negros Occidental, from Bacolod to Sagay, is sugarlands. Except for the roadside ditches and orange-blossomed acacia, the sugarcane stretches from the highway clear to the horizon. The dense fields are forbidding, and the tall plants nearly block the sky. The National Highway leads north through this landscape of massive fields, passing tiny houses and plots of rice, occasional towns, and great sugar mills, where the cane is ground. The centrifugal mills, called centrals, offer the only sign of industrialization. Most land is devoted to the sugar haciendas, which have been owned by the same powerful families for generations.

This concentration of land has brought social and political unrest, as it has throughout the Philippines, where conflicts over land have marked rural areas for hundreds of years. The lack of cropland for small farmers has brought widespread poverty; it is the main reason that millions have become migrants; and it also lies at the root of many of the archipelago's environmental problems. It is not possible to understand the misuse of the country's natural resources without understanding the patterns of land use and the reasons that they have endured.

In Korea, Taiwan, and Japan, programs that redistributed land and reformed tenancy arrangements after World War II laid the foundation for later economic prosperity. While similar undertakings have been repeatedly proposed as a partial solution to the landlessness and poverty in the rural Philippines, they have had limited scope and success. The most recent such program, the Comprehensive Agrarian Reform Program (CARP), was created in 1988 under the administration of Corazon (Cory) Aquino to address the inequitable access to land. The program set a goal of distributing to about 4 million peasants nearly 25 million acres of land, owned both privately and publicly. The program would also provide support services for farmers and establish a leasehold system to improve relations between tenants and landowners.[1]

CARP was originally intended to end in 1998, but after considerable public pressure, Congress extended it for another 10 years. By 2002, 1.1 million farmers had benefited from leaseholds covering more than 3.5 million acres of land. About 5 million acres targeted for distribution had not yet changed hands. Most public lands that fell under CARP had been fairly easy to handle, but the Department of Agrarian Reform (DAR) had succeeded in distributing only about half of the private lands. In fact, less than 15 percent of private lands that were to be acquired through what was called compulsory acquisition—as opposed to land that was offered voluntarily—had actually been distributed. In addition, many of the farmers who potentially could benefit from the program knew little about what it offered. One recent DAR study found that about one-third of farmers interviewed did not know that CARP existed.

Unlike the earlier agrarian reform programs in the region, CARP was created within a democracy. It also was created within a political context—both national and international—that offered little support for land redistribution.[2] CARP encountered many obstacles, including inefficiencies and corruption within its own institutions; early leadership at the DAR was implicated in land scandals.[3] The law itself has been controversial, and the program has lacked adequate funds. Landlords put up a great deal of resistance, with tactics ranging from legal wrangling to violence. They subdivided land, "converted" it to non-farming purposes, and generally refused to cooperate. Because CARP made land available for peasants to purchase rather than distributing it outright, there also were problems with their being able to afford and maintain their hard-won fields.

Not long before CARP was extended for another 10 years, I visited Negros Occidental, which offers a vivid example of the obstacles that agrarian reform has faced. The province's record for land distribution is notably poor; after nearly a decade, only about 32 percent of land originally targeted by CARP had changed hands, mainly because of resistance from the landowners. The haciendas abutting the National Highway, which were not distributed because of this opposition, are emblematic of the situation there.[4] With sugar haciendas covering so much of the province, most of the rural population is landless and extremely poor. Although sugar has at times been a lucrative industry, workers in the industry typically have earned the least of all agricultural laborers, and with few other options, during the periodic downturns they suffer. In the

1970s and 1980s, when sugar prices plummeted on the global market, nearly one-third of all sugarlands were left unplanted, bringing widespread starvation. A 1982 study of more than 170,000 children on Negros found that nearly 70 percent were malnourished. In the mid-1990s, falling sugar prices threatened a return to conditions of near-famine.[5]

I was directed to Negros Occidental by Victor Gerardo (Gerry) Bulatao, a DAR undersecretary with roots in the anti-Marcos opposition and the NGO sector. Gerry also sent me to Teodorico (Teody) Peña, a former provincial officer in the DAR who headed Quedan Kaisahan, a nongovernmental organization (NGO) in the provincial capital of Bacolod that assists farmers and farmworkers trying to obtain land under CARP. Teody invited me to accompany him to the northeastern corner of Negros Occidental, where a land dispute was unfolding in the village of Baviera, Sagay. A major landowner who had been arranging through CARP to sell land to the farmers living there had abruptly withdrawn his offer and instead threatened to evict them. Despite the local drama, Baviera's conflict had not reached the pages of the national papers. It was just one of many conflicts stalling agrarian reform, just one more land dispute, significant precisely because it was so typical.

Kaisahan means "solidarity," and *quedan*, "nameless," Gerry said as we spoke about Negros in his office one morning: "Because of the politics there, it's better to be nameless." Teody had been pushed out of office because of those politics. A stocky man of medium height, he was, even when I first spoke with him on the phone in Manila, outgoing and forthright, with little of the initial reserve that I usually encountered in the provinces. In Baviera, he would be meeting with farmers who were looking for counsel from Quedan Kaisahan. Joining us on the trip was the organization's head of policy research and advocacy, a thoughtful, broad-faced woman introduced as Glenda. The local DAR representative—the municipal agrarian reform officer—who was working closely with the farmers to resolve the dispute would also meet us in Baviera.

Early in the morning, we headed north on the National Highway. A surprisingly good road, the highway was maintained to bear the weight of trucks going to and from the centrals. There were few other roads. Sugar so dominated that part of the province that in some towns the cane covered as much as 80 percent of all agricultural land. We occasionally saw clusters of homes with banana trees, vegetable plots, and fields of flooded rice, but only on slivers of land squeezed among the haciendas. We also passed several open areas devoted to housing fighting roosters, whose bloody matches with sharpened beaks were a focus of avid betting. These fields—neatly lined with dozens of small, roofed shelters, each with a plumed cock tethered to a stake—were larger than the few rice fields we passed. Elsewhere I noticed growing at the edges of the cane fields a few vegetable plants: *camote*—sweet potato—and *kang kong* or other greens. On one stretch of road, an old man bent over beside the highway, picking the arrow-shaped leaves of *kang kong*, the food of the poor.

Under CARP, agrarian reform communities were being organized as a way to assist farmworkers in their efforts to buy and manage land. In Baviera and other areas, these communities gave farmers access to credit, technical aid, and other government services. Although such services were known to make it more likely for new farmers to succeed, the residents did not have access to them in that part of the province. "I was so scandalized at the greed when I came here," Teody said one night. The farmworkers could not get the government assistance, he said, because they "couldn't get roads. They"—the sugar planters and millers—"wouldn't give right-of-way for farmers to make roads."

Poverty was visible in Negros Occidental—in people's homes, faces, and bodies. Because the poor worked in the fields, their bodies were muscled and lean, but their narrow faces showed their hunger. The farther we traveled from Bacolod, the thinner people became, until we reached Baviera, where faces were so hollowed that cheekbones seemed about to poke through skin. Even the eyes of the children looked tense and defeated. After seeing Baviera, all the other faces I saw—including those of Glenda and Teody—looked healthy and full. For weeks after I returned to Manila, no one's face looked the same.

• • •

Land has been the root of unrest in the Philippines for hundreds of years. Conflicts over land sparked repeated rural uprisings against the Spaniards, as they took more and more cropland for their own, and it fueled the discontent that eventually coalesced into the 1896 revolution. It united peasants in the Huk rebellion of the 1950s, which led to the formation of the Communist Party of the Philippines; its armed wing, the New People's Army (NPA); and the call that land be returned to the tiller. Despite the deepening poverty and economic stagnation—and the social unrest that they brought—the government and private sector repeatedly failed to make rural economic development a priority. More than half of all rural families continue to live below the country's official poverty line, and they account for the majority of the poor.

The country's growing population contributes to the landlessness among the peasants, as does a terrain so mountainous that less than half of the country's total area is considered arable. A main cause of landlessness, however, is the highly skewed landownership. Nearly three-fourths of rural households have no land or little land, and an estimated 5 percent of rural families—and likely as few as 1 percent—typically have owned more than 80 percent of all farmland.[6]

Clearing land to farm.
BARBARA GOLDOFTAS

The inequality is probably more pronounced than agrarian reform advocates have realized. As with statistics about logging and fishing, there are no figures to describe adequately either the landlessness or the concentration of land ownership. Census data, for example, count only the number of farms and their size, without revealing whether landlords own farms in different places, as many do, or whether individual holdings belong to those of a large landowning family. These data appear to show that two-thirds of all farms are smaller than 14 acres, which would imply a relatively even distribution of land. Even casual observation shows that, as in Negros, plantations are far vaster than 14 acres. According to James Putzel, who studied agrarian reform in the Philippines, the census data, which landlords and some officials have used to rebut calls for reforms, were "designed to disguise land ownership."[7]

The extent of landlessness is even harder to assess. The figures usually cited by the government and NGOs range from about half of rural households to as high as 85 percent. Many of the landless are farmworkers, whose wages are among the lowest in the country. The landless include tenant farmers or sharecroppers, who pay as much as one-fourth to one-half of their harvest to rent and farm a small plot. They also include those who are even harder to count: the millions of migrants, including farmers tilling marginal land in the uplands.

The concentration of land ownership varies by region and by crop. While rice in Central Luzon tends to be grown by tenant farmers in fields of a few acres, other crops such as sugarcane are cultivated on the vast haciendas. The sugar haciendas generally are worked by laborers rather than sharecroppers, and of all crops, sugar has had the most unequal land holdings, particularly in Negros Occidental.[8]

       ◆   ◆   ◆

After traveling for several hours, Teody, Glenda, and I arrived at the community center in Baviera. Built as an addition to the house of the community organizer, a reed-thin man introduced as Efraim, the center consisted mainly of a corrugated metal roof and a *sari-sari* store. When we drove up, about a dozen people were resting there on benches and stones. Many of them had walked an hour or more to tell Quedan Kaisahan about their situation.

Baviera is one of two dozen *barangays* in Sagay, which then was a town of about 131,000 with two sugar centrals and an economy that revolved almost entirely around sugar. Like the rest of Sagay, Baviera was mainly haciendas. Where the cane fields ended, a cliff rose, its soil dry and crumbly. This was Mine Site, part of the area under dispute. Mine Site belonged to the Hofileña estate, which covered about 1,570 acres, or about one-third of the town. The Hofileña family was among the largest landowners in Baviera, which was named after the family that donated the site for the *barangay* in 1931. Steep and far from ideal for farming, Mine Site was the kind of land that people till when they have no other choice. It hardly looked worth fighting over.

The land in Baviera had not always been farmed. A silica mine opened there in the 1960s—hence the name Mine Site—but closed in the mid-1970s when the silica market dried up. By that time, Baviera had become a stronghold of the New People's Army, which has deep roots in Negros Occidental and other extremely poor rural areas. During the years that the NPA was active there, the communities abandoned much of the area. The hills planted with cane were taken out of cultivation when prices dropped during the 1980s sugar crisis. Hearing about the untended land, families began to move there and clear land to farm. By the mid-1980s, there were about 400 households on the site.

The enactment of CARP gave the farmers several options. Unlike previous programs, CARP covered all crops, and all land being farmed was considered productive, that is, "CARP-able." DAR would not accept undeveloped uplands with a slope greater than 18 degrees, which by law were supposed to remain forested. After being surveyed and titled, each parcel of land had to be approved by the DAR, which would seek landless farmers, called "farmer-beneficiaries," able to buy it. Once a parcel of land reached this stage, the Land Bank of the Philippines would assess its worth, based in part on its productivity; make an offer to the landowners, who could reject it; and pay them "just compensation." Farmers could then buy the land, making their payments over 30 years.

CARP also allowed for the creation of leasehold contracts, arrangements that the farmers in Baviera made through the municipal agrarian reform officer. When farmers had previously been considered squatters, a leasehold would set such conditions as a rent for the land and the percentage of the profits to be paid to the landowner, usually about 25 percent of average net production. The leasehold agreement could be a preliminary step toward buying the land, giving the farmers a small measure of security. Stable access to land—security of tenure—was unusual for the millions of migrant farmers. It was also believed to be an important incentive that would motivate people to farm their land more sustainably.

During its initial years, CARP focused mainly on land that could be distributed easily: public land, private lands that were idle or abandoned, and land that owners offered for sale voluntarily. Only later would the government require that owners sell part of their holdings through compulsory acquisition. A CARP handbook explained that including private land was "inevitable" because of the "sheer extent of landlessness."[9] Requiring landowners to sell proved difficult, though, and in Negros Occidental, DAR's failure to acquire land from the major landowners encouraged other landowners to resist as well.

The town of Sagay was typical of the province. About 69,000 of its 84,000 acres were agricultural land, mostly sugarland, of which only about 14,800 acres could easily be distributed under CARP. Michael Salcedo, the municipal agrarian reform officer, called those hectares "workable CARP-able." Between 1988 and the late 1990s, about 14 percent of those fields had actually been distributed.

There had been no compulsory acquisitions in Sagay, only land offered at the landowner's initiative. In the early 1990s, three of Baviera's main owners made voluntary offers for sale, including one for about 990 acres or about two-thirds of the Hofileña family's holdings. Mine Site, which the landowner later withdrew, was among them, although it had been secured—or so the farmers thought—with a leasehold contract. About the same time, the landowner secretly filed for a permit to start mining again, one way to gain exemption from agrarian reform.

◆ ◆ ◆

As is customary in rural villages, the meeting with Quedan Kaisahan began with a few words from the *barangay* captain, who had first been elected in 1972, the year Marcos declared martial law and suspended elections. He turned the meeting over to Efraim, the community organizer, who recounted the events that had led to the meeting. Some months earlier, the farmers had unexpectedly received notice, from posters tacked to trees, that the voluntary offer for sale (VOS) had been halted. One of the hand-painted signs read: "*Aviso,*" the Spanish word for warning. "No one should take possession of the land in Manhill since VOS has been stopped. Effective 1-1-97. Order by Landowner."

Once land has been offered for sale voluntarily, it cannot be withdrawn, but the farmers did not know this. "People came to me," said Efraim, "to tell me that there was a sign that prohibits them from cultivating the land. The owner had withdrawn the letter of intent to sell the land. That was about five months ago, in March, just before the April elections." A number of the farmers laughed; financial transactions and violence often preceded elections. "Later, some people, especially in the mine area, were given a letter telling them to evacuate the area," Efraim continued: "The overseer brought me the letters. The letters ordered them to stop cultivating their farms: 'If you don't leave, your house will be demolished by bulldozer.' There was no [other] warning. No one moved. Until now, nothing has happened."

He spoke gravely: "We have been waiting to hear, but there have been no other actions. We're not at peace with ourselves. We still worry about what will happen, worry that the next move of the landowner will be more drastic than the earlier warning. The landowner's daughter came last month and warned that those who couldn't pay the lease payment for the rest of the year, who fell behind in their payments, would be ejected. We had never seen her before."

After Efraim finished, others told their own stories. The small circle contained much of what made land disputes throughout the country so wrenching and contentious. These were people who worked hard, often laboring in the sugarlands to supplement what they earned from their own small farms. They came from different backgrounds, and some had ties with the New People's Army. Their education varied; one man, Teody said later, did not understand my simple questions. The farmers had few options, and many had lived through the faminelike conditions of the 1980s. Some worried aloud that their family would "go hungry" if they lost their farms. Others were more restrained. The government officials seemed concerned; they seemed competent and sincere. The NGO workers were experienced and worked well with the government. The farmers were organized. Under CARP, it should have been possible for them to buy the land. They likely would not have the chance.

❦

During most of the Spanish era, which lasted until the late 1800s, the Philippines was isolated from the rest of the world, its ports sealed to all countries but China and Mexico. The Spanish galleons charting the route between Acapulco and Manila were once the largest ships in the world. Even as late as 1818, the most lucrative export leaving the Philippines was birds' nests, plucked laboriously from limestone cliffs. The port of Manila, which was closed in 1565, opened to outside trade shortly before 1800, and in 1855, Iloilo City, not far from western Negros, became the country's second foreign trade port. The new trade spurred the development of export agriculture and the expansion of haciendas for sugarcane, which until then had been grown in small plots, as the crop still is in some countries today.[10]

According to conflicting stories, sugar was introduced to Negros in the 1850s by either a Frenchman from Normandy or a young British vice-consul who, driven by visions of sugar exports, transformed the forested lowlands into haciendas of tall cane.[11] The climate and soil proved ideal for sugar, which flourished and, as prices stayed high for several decades, became one of the country's four leading crops.

Until the mid-1800s, the island of Negros was still a single province, forested and undeveloped. Although often depicted as unpopulated, established groups of Malay farmers lived in the lowlands and Negrito foragers lived in the uplands. Within about 40 years, sugar planters had arrived, cleared the forests, and planted haciendas. The planters recruited seasonal workers from Iloilo and Panay, then the second most populated province. The natives who refused to become workers on the haciendas either fled into the uplands or were murdered.[12]

The sugar work was difficult, and to force workers to toil, landowners relied on physical cruelty and often bound them to the plantations with debt.[13] More than 150 years later, sugar work remains arduous. Like many sweet plants, sugarcane protects itself, and the sharp edges of its narrow leaves can rip skin. The work is also intermittent. Between May, when one crop is cut, and September, when the next is planted, workers must endure the "dead season," generally surviving on rice provided by the planters, who deduct its cost from the next season's paychecks. During harvest, the dense, succulent cane is hard to fell and heavy to haul, and afterward, the stalks that remain are burned, leaving sticky, soot-covered fields to be cleaned and plowed for the next crop. In *Dead Season*, a tale of violence and feudalism in Negros, writer Alan Berlow describes a burned cane field as "an endless prison of blackened upright bars, each oozing a thick, viscous gum and covered with a fine silken ash. Men working in burned fields swaddle their heads in towels or rags and the rest of their bodies in thick denim, leaving only their faces exposed. At the end of the day they emerge from the fields looking like overgrown tar babies."[14]

The history of sugar in Negros Occidental is a history of far more than one industry. Beginning in the early 1900s, the planters received preferential treatment in trade with the United States, and during the 1920s, as the country's economy grew ever more linked to that of its colonial power, sugarcane increasingly was grown for that one export market. To meet the rising U.S. demand, the sugar barons built the first centrals, replacing the smaller and less efficient light steam mills and further fueling the expansion of the haciendas. As the planters benefited from the high prices and preferential relationship in U.S. trade, they expanded production and by the 1950s were producing more than half of the country's total sugar output. In Negros Occidental, they repeatedly devoted their profits to expanding their own wealth and power rather than developing the industry, which could have benefited the province as a whole. Instead of investing in new technology to intensify production, for example, they acquired more land, which brought them wealth as well as greater prestige and political power. They also failed to raise workers' wages; doing so could have raised the standard of living in the province but would have made the workers less dependent on the haciendas.

The growing stature of the sugar barons in Negros Occidental in time had an effect that rippled throughout the country. This elite has produced a disproportionate number of national politicians. They are among the major landowners, and they form a small epicenter of political and economic power. Even in the 1990s, more than two-thirds of the major landowning families that dominated the manufacturing sector could be traced to the sugar industry, and many of them made their fortunes in Negros. According to the political scientist Temario Rivera, the power of the sugar families has constrained the Philippine economy which, despite its strength in the 1950s and 1960s, failed to keep pace with those of its neighbors in the region. In part, widespread poverty limited the development of a middle class and a strong domestic market for manufactured goods and other products. In addition, ties that leaders in manufacturing had to the land and to such industries as logging, mining, and agriculture—especially sugar—created conflicting interests within the elite that undermined their commitment to manufacturing and further obstructed the country's industrial development.[15]

. . .

In the early 1960s, the bishop in Bacolod, Antonio Yapsutco Fortich, was one of the first to condemn the planters for mistreating their workers. Although from a wealthy sugar family himself, he is well known for having described Negros as "a social volcano" that was just "waiting to explode."[16] Fortich's outspoken criticism of the industry continued for decades, and, in an effort to unionize workers and ease conditions in the industry, the Catholic Church helped form the National Federation of Sugar Workers (NFSW). Despite pressure from the

NFSW and the New People's Army, poverty among the sugar workers persisted. Although the island did not "explode," decades later conditions have not improved, not in Baviera or elsewhere.

At the meeting in Baviera, one of the first farmers to speak was a 35-year-old mother of five, Teresa, who had walked an hour to get there. Teresa still worked on the haciendas. "Before, I stayed in the *barangay* proper. When I got married, we transferred to Mine Site, in 1985. We farm about two hectares [about five acres] in sugarcane, corn, cassava, and *gulay* [vegetables]. We don't earn enough. I also work in other sugarcane areas. I plant and my husband plows. We work all year long if there's work available. But that doesn't mean we work seven days a week. The landowner will tell us if he has some work to be done. We're employed by small planters, and they come themselves to tell us if they have work. We earn about 50 pesos a day [about two dollars]."

"Sometimes we're hired on a daily basis—for 50 pesos—and sometimes on a piecework basis," she said. "If we weed, we're paid 400 to 500 pesos [about 16 to 20 dollars] per hectare. It might take three weeks for my husband and me to weed the field, working eight hours a day." Over three weeks, their daily rate would bring them each 750 pesos.

It turned out that even their daily rate, which they abandoned for the sake of steady work, fell far below the legal limit. The minimum wage for sugar workers was 90 to 100 pesos (about 4 dollars) a day. Over three weeks, the minimum wage would bring 1,500 pesos, or about 50 dollars each, twice what they actually earned. Teresa did not dwell on the unfairness of her situation or its illegality. "Even if we're paid less than that," she said briefly, "we are still forced to do the work so we can feed our family."

After the meeting, Teresa and another woman, a former NPA, joined us for lunch. Shy yet unafraid of appearing hungry, they sat with their spines curled against their chair backs and ate plate after plate of rice, servings twice those served in the city, onto which they scooped spoonfuls of *ulam*—viand or any food eaten with rice. During the meeting, Teresa had smiled frequently as she spoke, especially when one of us looked at her. On the return trip, Teody and Glenda commented that she had been extremely nervous, especially when she spoke about the landowner's threat. "She almost cried," said Teody. "She seemed as though she was reliving it."

"I'd seen the signboard in March," said Teresa. "I was saddened by the notice. I stayed there for a long time, and I was thinking that now I will be evicted. We received a letter on April 2. When I got the letter, I was afraid because my house would be demolished. I referred the letter to Efraim and asked for his help. I talked with my neighbors, and we invited Efraim to have a meeting with us at the Mine Site. It's hard for us to act individually."

Another farmer, 39-year-old Rene, had moved from southern Negros to Baviera in 1989 because he had a relative there. A small, lively man with deep smile lines and a mustache, he had a farm of about five acres, steep land where he grew bananas, corn, and sugarcane. The father of two children, he admitted, "I am afraid to have more children because of the insecurity of the land. I'm not certain that there won't be more threats." Like the others, he had seen the signboard nailed to the tree, and he also received a letter. "I felt helpless. I want to continue tilling the land—that's the only source of income I have. I want to keep the land through legal means."

Under the guidance of the municipal agrarian reform office, Rene and others had entered into a leasehold contract and paid the lease rental to the office, which kept their funds in escrow. Although some others had not started paying yet, Rene had been doing so since 1996. "I'm supposed to pay 750 pesos [30 dollars]," he said. "I've paid more than 200 pesos, will pay the rest by December."

Rene claimed to have finished high school, which Teody questioned later, and he spoke with pride about his life: "After I finished high school, I wanted to look for a job that could support my family. I worked as a forest guard for eight years for the DENR in the research division. The research was transferred to Cebu, so I moved here to find work. I heard that there was land, so I cleared some. I never asked any permission, I just cleared the land. I had no experience farming; I learned from practice. I learned by doing.

"I tried to raise my family through legal means," he said. "I planted corn and bananas because my family needs them—and *camote* [root crops] and trees. I planted sugar because I thought it would bring enough income to sell to the sugar central. They buy it and mill it. But they bought it at a low price. I sell the root crops, the vegetables, everything I grow. It's not enough for the family. All I want is the security of tenure, and I'm willing to pay for it."

. . .

The basic patterns by which land is owned and used in the Philippines can be traced to its colonization by both the Spaniards and the Americans. Early in the 1500s, before the Spanish conquest, people lived in *barangay* communities beside coastal waters, rivers, and mountain streams. Boats were the only form of transportation, and houses were built in a line without roads between them or between neighboring *barangays*. While land is usually described as having been communally owned before the Spaniards arrived, the economic historian O. D. Corpuz asserts that there was actually a defined system of land ownership and tenure. Families owned the lot on which their house stood, and behind the houses lay family fields, often in strips. There usually was, in addition, a community-owned tract covering the nearby forests, uplands, fisheries, and coasts.[17]

Over a period of about 150 years, the Spaniards brutally conquered the *barangays*, one by one, resettling them into pueblos with a church, plaza, town hall, and houses that, instead of following a line, filled a grid of streets. As before, each family had access to its own lot, a nearby field, and common land. The difference was that the king technically owned all the land, leaving the *indios* with the right to use land but not own it. The king also awarded tracts of land to the powerful church and religious orders. These estates, usually run by tenant farmers, formed the basis of what came to be called the friar lands.[18]

The lands held by the religious orders were far larger than the common *pueblo* lands, and by the early 1700s—or even, by some accounts, the early 1600s—the powerful friars had begun to expand their holdings illegally, accumulating vast estates, parcel by parcel. Corpuz describes "bulky bundles of eighteenth century handwritten records of hearings, statements and other reports of proceedings in land disputes" that tell the story of the expansion of the friar estates and the anger and unrest that the "land-grabbing" sparked. By the late 1800s, the friar orders held close to half a million acres, more than half of it near Manila, where they grew rice, fruit, vegetables, and other crops to sell in the city's markets.[19]

The U.S. Philippine Commission, realizing that discontent over the friar lands helped spark the 1896 revolution, claimed that dividing up the extensive estates would be a priority. Although the colonial government did buy some of these estates, little land was sold to tenant farmers. The landed families—many of whom had cooperated with the Spaniards and would later do so with the Japanese during the World War II occupation—also cooperated with their new colonizers. The Americans, rather than undermining the landed elite, chose to support it.[20] Most of the land purchased from the friars was sold to large landowners, thereby perpetuating the inequality and landlessness and creating, among the landowners, an enduring opposition to land reform.

To address the plight of the growing numbers of landless farmers, the Americans also pursued a plan to resettle them in the frontier of Mindanao. This approach, based on the homesteading that settled the U.S. West, did not succeed; fewer than 10,000 peasants set up homesteads, some of them merely schemes to get landowners more land. In 1936, a year after the United States granted the archipelago its commonwealth status, a survey of rural areas reported "complete penury" among tenant farmers.[21]

In the aftermath of World War II, the land reform programs that the Americans supported in Taiwan, South Korea, and Japan distributed from 27 to 40 percent of the cultivated land and affected between 63 and 76 percent of all agricultural households. This redistribution of resources has been credited with helping to break the feudal landlord power, build a strong middle class, and lay the foundation for future prosperity. In the Philippines, the United States took a different approach and one more reflective of its subsequent behavior as a colonial power, responding to the persistent rural unrest in the early 1950s with military action and economic and technical aid.

After declaring martial law in 1972, Ferdinand Marcos pronounced the entire country to be under land reform, but the programs mainly covered rice fields. They also accomplished little. Rather than addressing the causes of the growing rural unrest, Marcos and his U.S. allies opted to quell it with a military response and general rural reforms. Despite the rapid economic growth of the 1970s, landlessness and the intense concentration of wealth deepened. There were few economic alternatives to agriculture, and the abject poverty in rural areas eventually deepened the pressure on the natural resources and weakened the larger economy.[22]

In 1986, the ousting of Marcos and the election of Cory Aquino brought hope that the political and economic turmoil that were crippling the country could be reversed. At first, there was no overt opposition to using agrarian reform to address the rural inequality. In 1988, even the largest landowner organization expressed a commitment to a "truly just, workable, and effective" agrarian reform program.[23]

As the widow of Benigno (Ninoy) Aquino, whose 1983 assassination helped propel the opposition against Marcos, Cory Aquino had been selected to run in part because she was a popular figure perceived as sympathetic to the poor. In a campaign speech about social reforms, she named agrarian reform as her first priority, citing its two distinct goals, "greater productivity" and the "equitable sharing of the benefits and ownership of land." These goals, she admitted, could "conflict with each other." Nonetheless, she promised, the government would seek "viable" solutions, taking into account the needs not only of agriculture, but also of those who worked the land. "For longtime settlers and share tenants," she explained, drawing on language more often attributed to the peasant movement, "land-to-the-tiller must become a reality, instead of an empty slogan."[24]

The broad alliance that brought Aquino to power, however, had differing views on land reform. Some supported what Aquino had called "genuine agrarian reform." Others claimed to be willing to share "the fruits" of production but not the ownership of the land.[25] Aquino herself came from the wealthy Cojuangco family, which owned the 16,000-acre Hacienda Luisita in Luzon. Within a few years of taking office, she would be bitterly criticized for having maintained her allegiance to her family and class and not pressed hard enough for agrarian reform.

During the initial months of her presidency, with the constitution suspended, Aquino wielded enormous power and without congressional approval could have implemented wide-ranging reforms to redistribute land. Instead, she chose to define agrarian reform as a program to enhance production through a "support system" of credit and other assistance to small farmers and incentives to large agricultural corporations to share their profits with farmworkers. She did not appoint a minister of agrarian reform for two months, and those she did appoint were inexperienced in rural affairs and did not last long in office.[26] She herself backpedaled on an apparent campaign promise to distribute some of Hacienda Luisita to its tenant farmers.

The turning point came in January 1987, not even a year after the election, when about 20,000 unarmed demonstrators led by the peasant leader Jaime Tadeo marched toward Malacañang—the presidential palace—to demand that land reform become a priority. At the Mendiola Bridge, they were met with barbed wire, police, and gunfire. Thirteen died and dozens were injured in what came to be called the "Mendiola Massacre."[27]

The Comprehensive Agrarian Reform Program was enacted the following year. The law has proved controversial. Unlike earlier laws, it covered all crops, and it included compulsory land acquisition. Having been hammered out in Congress mainly by traditional politicians from the elite, however, it contained, critics contend, provisions that favored landowners, agricultural corporations, and the more powerful families.[28] Unlike earlier programs in the region, it required the government to pay for the land (with "just compensation"); the farmer-beneficiaries in turn would have to buy the land from the government. Landowning families could keep about 12 acres per individual and heir over 15 years, giving them the latitude to disguise their extensive holdings by transferring portions to relatives and others. Despite these provisions, because of the uncertainty of what CARP might bring, some landowners stopped investing in their land, limiting both production and employment.

The agrarian reform process itself has been complex. Multiple agencies had to be involved with each case, and the two with the most responsibility, the DAR and Land Bank, were inefficient institutions run by Marcos-era staff. Bureaucratic snafus—such as parcels of land that lacked titles, acceptable deeds, or proper surveys—snarled the process. In some cases, the land had not been assessed, or its assessment had been challenged, or Land Bank would not buy the land because it was not being farmed and therefore was not considered productive.

After the meeting in Baviera, Michael Salcedo, the municipal agrarian reform officer there, described the problems in "processing" cases: "There is a slow and *low* evaluation of the land by Land Bank so that it's not worthwhile for the landowner to sell." Land Bank often will appraise land for, on average, 4,000 pesos a hectare for rice and 2,000 pesos for corn, basing their figure on the average productivity in the area. Over the past year, sugarlands were valued at 114,000 pesos a hectare, he said, "so the low valuation makes VOS [voluntary offers for sale] not worthwhile."

Some DAR officials opposed land reform or were not aggressive enough. Teody spoke about them with his characteristic bluntness. "What did they expect, that the landowners wouldn't resist? You're taking somebody's land. But when they resist, are you willing to fight back, are you ready to place the land under CARP, no matter how strong the resistance? One thing the DAR lacks is creativity in entering the land. Land Bank is also too slow. They don't have the staff. But they were given the responsibility under the law."

The provincial DAR office was an old building in Bacolod that was not well kept up; in the center courtyard, water from the night's rain dried slowly in the morning air. Stephen Leonides, the provincial agrarian reform officer, or PARO, spoke matter-of-factly about the cases that passed through his office. On the white board behind his desk, he had drawn two triangles, one inverted, one upright. A line divided each close to its apex, with the number 10 written in the small section and 90 in the large one. He pointed to them. "This is how Negros Occidental compares with other provinces. Ten percent of the landowners in the province own 90 percent of the land. Instead of sharing some part of the land, they keep it themselves. That is the culture here."

In 1996, the annual target for the provincial DAR—based on the land remaining from CARP's original goal—was 22,000 hectares. They were able to distribute 5,000. The target for 1997 was 23,000. "We've distributed 1,412 so far," said Stephen. Having worked for the DAR since 1983, he presumably had seen far worse. "There are nearly 17,000 hectares in the pipeline. What are the main bottlenecks? For nearly 11,000 hectares, there isn't landowner resistance, just bureaucratic slowness. Of those, 6,000 are under process at this office."

In Negros Occidental, the landowners' resistance was legendary. They responded slowly to bureaucratic requests or failed to submit the necessary documents. They also hired security guards or private armies to prevent anyone from entering their property, even to survey or assess the land. In some cases, they would draw on powerful politicians to pressure the DAR for exemptions. They would also target local government workers. "Once they receive notice that their land has been covered by DAR," said Stephen, "they file harassment cases against DAR personnel, administrative cases. The most difficult kind of resistance is the harassment cases. There's not a high turnover, but sometimes the MAROs [municipal agrarian reform officers] get demoralized."

Harvesting sugarcane. Outside Bacolod, Negros Occidental.
MARISSA ROTH

As the 1998 deadline approached, it had become even more difficult to distribute land, Stephen added. "It was easier to implement CARP earlier because the landowners were afraid to go to the fields. There was less resistance. There's more resistance now, because the landowners think that CARP will end next year." There was no sure way to counter the resistance. The DAR office might work with the local people's organizations, which could negotiate with the landowners or help them enter the property, or with the Philippine National Police, although, as elsewhere, they were unreliable. "We can't always get the full support from the military," Stephen said. "Some detachments are supported by the landowners—who provide their food and supplies—and they're based in the haciendas. So even if we have a memorandum of understanding with the police, we do not expect them to cooperate with us. Sometimes when we distribute the land, we ask the Philippine National Police [PNP] to accompany us. We arrive, and then the landowner arrives with his PNP. So rather than distributing the land, we end up negotiating."

Books and brochures about CARP portray the farmers as smiling peasants with bright eyes. That was not the reality in Baviera. The province overall was polarized ideologically, with a history of violence and militancy by both the left and right.[29] The New People's Army called for land reform, and the National Federation of Sugar Workers called strikes. Some sugar workers were active in the peasant and underground movements. "People sometimes burn the sugar-cane of the landowner," said Stephen. "We have lots of voluntary offers for sale because of that strategy." He laughed. "That's done in the south of the province, not as much here."

The relationship between the landowners and sugar workers was marked by dependence as well as peasant resistant. Some sugar workers remained so dependent on their landowners that, out of fear of retaliation, they would not organize or even stand up to them, which further slowed CARP. Michael Salcedo's office worked closely with the cases, preparing documents; keeping track of claims in thick, legal-sized folders; and handling land disputes. In nine years, the DAR had only distributed about 19 acres through compulsory acquisition. "In that case, the landowner was no longer in possession," said Michael. "The farmworkers had taken the land already. That doesn't happen here much. The farmworkers are afraid."

"In the south of the province," Michael said, "the farmworkers are well organized. When the leader says, 'We won't work,' the haciendas will be paralyzed. Here, we're still in the preparatory stages. People are quite aware of their rights, but they're still afraid. They have to feed their families. Most of them are dependent on the landowner. They will ask me, 'If we submit a proposal, can you let our children go to school, can you pay for our hospital bills?' They depend on the landowners for all this." One peasant organizer in Negros Occidental described how some sugar workers "literally cry upon seeing the DAR people in their community. They don't want to lose the patronage of their landowners."[30]

When they interviewed potential beneficiaries, said Michael, many would not even talk with them. "They worry that if they attend the meeting, they won't have work by the next week. When there are forms to be filled out, the farmworkers are hesitant to fill them out. And it's not just their worry, it's what actually happens. Once after we conducted an information drive on Friday, on Monday some farmer-beneficiaries were no longer hired to work. We negotiated with the overseer and told the landowner, and they got their jobs back."

Farmers' dependence could continue after land was transferred. One of the landowners' major arguments against agrarian reform has been that farmworkers were not qualified to run farms and that they often did not even want the responsibility. It was true that, in Negros Occidental, some farmer-beneficiaries had failed to keep up with their payments or could not manage their farms. So far, many farmers who gained land under the program had either sold the land or, unable to make the payments, had lost their farms. In part, this was because of the way that their payments were calculated. They had 30 years to pay for their land, but their payments were based on their gross production—what they might earn from the harvest—without taking into account their costs.

The DAR found that farmer-beneficiaries were far more likely to succeed when they were grouped into cooperatives—the agrarian reform communities—and received the support of government services, such as those that could not reach the farmers in Baviera. Even so, Michael acknowledged, "some farmer-beneficiaries sell or lease their property to other farmer-beneficiaries. They're used to getting a salary. Perhaps they need money right away, to pay for hospital expenses or whatever. This happens more when they are not members of cooperatives. Our government here is very lax. There's no punishment for farmer-beneficiaries who sell their land."

. . .

One night, Teody and I sat and talked on the patio outside the Quedan Kaisahan office. He had an unrestrained laugh and constant stories and jokes, but none of it masked his seriousness. As he drank beer after beer, he told me about a farmer he had helped during a land dispute. Encouraged by Teody, the man had stood his ground with the landowner, and he had been gunned down. Afterward, Teody recalled, the grieving wife came to him. "We believed in you," she said.

The death upset him deeply, Teody said quietly, and since then he has never pushed people to take more of a risk than they want to. "I have a way out," he said. "I can call Manila and get money. I can call California and get money. These people are here to stay. They're the ones who will get shot; they're the ones who will die. They have no way out."

In places like Negros Occidental, where conflicts are so polarized, it is all too easy to describe a process such as land reform in simple terms, blaming one group and glorifying the other. The reality was not that simple. When NGOs had an adversarial relationship with the government, they usually criticized its operations sharply, but former NGO officials who moved into the government often presented a subtler view of its strengths and failings. In a talk that Undersecretary Gerry Bulatao gave in between two stints at the DAR, he described the range of DAR employees he had worked with: "The state is not monolithic. The majority of civil servants I met at the DAR were honest and well intentioned, but not creative, persistent and dynamic. There was also a tendency on the part of many to be 'minimalist'—to do what is required to get by. . . . There were employees, it is true, who joined DAR or remained in DAR to look after their own interests—to complete a VOS, to get a higher value for their land or make money illegally by 'selling' land use conversion orders or other administrative decisions. But these people were the exception, not the rule."[31]

The landowners' resistance clearly had limited agrarian reform in the province, but so had the history of dependence, underdevelopment, and violence. In Baviera, the military conflict may have receded, but its residue remained, adding one more layer of complications that could divert a bureaucratic transaction.

At the Land Bank office, I met Antonio, who had worked there for nine years and was a well-meaning employee like those that Gerry described. His work used to take him into the field, and the reason he shifted to office work showed another perspective on what can slow down a bureaucracy's machinery. A slight man, Antonio seemed a dependable and capable worker who would focus carefully on technical details. As we spoke of the kinks in the Land Bank's process of assessing and valuing land, he suddenly began a story, without telling me why.

"I went to Baviera in 1990 or 1991," he said. "Inspecting the property was risky. There were instructions not to go there because it was NPA infested, and you couldn't enter the area without getting clearance from the military or police. When we went to Baviera, though, I felt safe—because I was with DAR and Land Bank representatives. We hiked in; it's safer to go by foot." These precautions were the very ones that I had been told to take when I first traveled in the rural provinces. In areas of military conflict, I was warned to enter with a local guide and never travel with the military, especially in their vehicles. Even those precautions would not necessarily alleviate suspicion, as Antonio had learned.

When he first interviewed the farmer-beneficiaries in Baviera, he remembered, part of the area was still classified as mineral land. The Land Bank could only purchase land known to be productive. He continued, "In some steep areas, I told them to cultivate the land. I was accompanied by the former overseer, who said the same thing. What if we pay and acquire the property and then find later on that the land is not CARP-able?

"I also excluded the portions along the creek and the portions that are hilly and probably couldn't be farmed." He paused, then commented about what he had just said, without giving a hint of where his story was headed. "We do feel that the farmer-beneficiaries are less fortunate and we should help them, but it's DAR that identifies the property, and we just value it. DAR and Land Bank play different roles. Land Bank has to make sound decisions as a bank; our major role is financial."

At a certain point that day, he left the others. "I came down from the mountain and went to a *nipa* hut to get a drink of water. I went alone. The owner told me to wait." He waited for what seemed an unusually long time. "After a while, the wife brought a pitcher and glasses, and we drank." Only after he returned did he find out what had caused the delay. "I heard that there had been an encounter in the next *barangay* the day before, and there were NPA inside the hut. I was wearing shorts and a cap and a brown shirt, and I had a camera at my side. The camera looked like a gun from far away. [The NPA] thought I was the military. Neighbors told me [later]. I was afraid and didn't sleep the whole night. So I remember Hofileña well. I don't do as much fieldwork anymore."

It was not uncommon for Filipinos who had come of age under martial law to tell me stories about the war that had marked their generation. I heard about brushes with danger, about arrests, about confinement in prison, time spent underground, fear. A man confided that to shorten his wife's prison term, during a conjugal visit they conceived their daughter. A woman who herself spent time in prison described how she and her fellow inmates hid her young daughter when the family came to visit, eventually triggering a policy that children be allowed to live with their mothers behind bars. People occasionally spoke of the infiltration of the NPA in Mindanao, the beheading of suspected "deep penetration agents" that followed, and the fear and disarray that the executions brought.

Usually I could tell the reason I was being told a story: The tellers wanted me to know them better, or they wanted the status that those experiences still brought. They wanted me to understand their hardships and the ways that the country's recent history reverberated powerfully in both political life and daily life. Although he did not say this, Antonio seemed to want to explain why he worked in the office rather than the field, but he also wanted me to understand how complicated a single case of land distribution could be, and why it was so easy for the Land Bank's work to stall.

"The Ilongo have a saying: Don't trust too much," Antonio said. "My wife told me that when I go to the field it's better to camouflage with others and not to care as much about wearing clothes that are comfortable. You never know what others are thinking about you. It's better if you go with someone who can identify you and if you're not all strangers to the place. We always take farmers with us. I have told my colleagues doing inspections to try to be extra careful and friendly to people around you. Once you're very friendly, you can eat with them, you can crack jokes with them. There's no problem." Antonio left to take a phone call, and when he returned, he adjusted his story slightly. "The NPAs were also friendly. I had talked with them sometimes. They were good conversationalists. They knew that the Land Bank is there to do good work, and they respected that."

    • • •

As it has for hundreds of years, land continues to lie at the root of economic and environmental problems in the rural provinces. Despite the gap that persists between the provinces and cities, the struggle for land has also changed urban life. As millions of people migrated within the country, fleeing landless-ness, militarization, and the effects of ecological decline, they moved into both sparsely populated areas and urban areas. Although the birthrate tends to be lower in cities than in the countryside, between 1980 and 1990, the urban popu-lation nearly doubled in size while that of rural areas remained the same.[32]

In cities, what might seem inevitable parts of life actually trace to the eco-logical degradation of the rural provinces. The deforestation of watersheds—around Cebu City, Puerto Princesa, and other urban areas—has diminished the cities' water supplies. In Manila, the growing population, including millions of squatters, adds to the pressure on the inadequate water and sewerage sys-tems. As the ecological chaos of the rural provinces continues to swell the cit-ies, bringing urban environmental problems with deep political and economic causes, potential solutions, like those for land use, will require far more than the passage of laws.

9 | Come What May: *Bahala Na*

In 1975, there were only five cities of more than 10 million people worldwide, mostly in industrialized countries. By 2000, there were nearly 20, mostly in developing countries. Urban theorists worried about these new megacities, portraying them as crowded, polluted, and unimaginably unlivable. Reports issued by international agencies described a coming urban explosion as chaotic masses of humanity spread uncontrollably, bringing political and economic instability and "frightening," even "life-threatening" environmental consequences. They warned that the condition of these megacities—whether they were prosperous or impoverished, peaceful or unstable, environmentally sound or not—might doom the stability of the entire world.[1]

Metropolitan Manila is one of those megacities. Made up of 17 separate towns and cities, by 2000, it was home to around 10 million people, 20 million including the surrounding areas. Hot and humid, dusty and crowded, it swelled from the ports and old medieval city into adjacent provinces, where new housing and industrial estates covered fields that for generations had been cultivated with rice. The metropolis had grown into a pastiche of old and new, poverty and affluence: luxury high-rises, marble-floored malls, and gleaming commercial centers imposed on congested old neighborhoods whose streets twisted into mazes. As modernity encroached on tradition, in a single block a fast-food franchise with picture windows might abut bamboo scaffolding bound by twine. In all but the most modern areas, which barred slow-moving vehicles, sleek cars with smoky windows crowded out pedicabs and tricycles trundling along the edges of narrow streets.

From the outside, a megacity such as Manila may appear bleak, but the capital is a vibrant cosmopolitan city, the country's intellectual and cultural center, its political core. Compared with the rural provinces, on average people living in Metro Manila have a far higher standard of living. They are better fed, they have better access to medical care, and they live longer. Manila also offers employment, a main reason that between 1970 and 1990 alone, millions of migrants surged into the city and the capital's population more than doubled. Manila continues to draw people. It is the place where people come to study, to find work, and, in the words of a woman living in a squatters' settlement, to "uplift their lives."

Manila's quick growth has not made it an easy place to live. In the 1990s, the growing population outstripped the roads, transportation, utilities—all the structures and services that form a city's foundation. For most people, living in Manila meant juggling daily the unreliability of this urban infrastructure. The water supply was intermittent and the sewerage system almost nonexistent. Garbage collection was inadequate, particularly outside of wealthy neighborhoods; waste accumulated in streets, blocking drains and waterways, and during the rainy season, many parts of the city flooded. The traffic was formidable. Rush hour took more than half of the day as streets teemed with pedestrians, pedicabs, jeepneys, tricycles, motorcycles, taxicabs, cars, buses, and even racing bicycles darting in and out. Epifano de los Santos Avenue, or EDSA, the swath of the highway built to ring the city that now serves as its spine, alone carried an estimated 2 million passengers a day.[2] Clouds of exhaust and other pollution left fine grit that covered walls, windows, waxy tropical leaves, hair and skin. The exhaust would sting the lungs, and phlegm expelled by the perpetual "Manila cough" was tinged with black. Traffic cops breathed through masks, and those who commuted by jeepney filtered the sooty air through folded handkerchiefs.

Manila's growing environmental problems took a toll on both public health and economic activity. The inadequate sewerage and water supply brought a high rate of waterborne disease among the poor. Flooding and brownouts caused lost hours and days of work and lowered productivity. The traffic, which only subsided late at night, slowed economic activity and was the main source of the air pollution that shrouded the city's streets, bringing high rates of respiratory disease and children's blood levels of lead.

As in the urban centers of other developing countries, the wealthy in Manila could partly insulate themselves from the environmental decline. On the congested roads, air-conditioned cars brought some relief from the traffic and pollution, and with a laptop computer, a chauffeured passenger could conduct business. The homes and workplaces of the wealthy had their own private infrastructure: cisterns, water pumps, back-up generators, even sewage treatment. Older, exclusive neighborhoods had a secluded feel, and walls topped with glass shards and barbed wire often enclosed the lush, manicured grounds of family compounds. The newer affluent neighborhoods were typically gated subdivisions. In fact, roads that were private or had limited access made up more than one-third of Manila's roads, which reduced traffic and freshened the air in these neighborhoods but also markedly aggravated the congestion elsewhere. In 1999, an official in the Department of Transportation and Communications admitted that, in order to fully solve Manila's traffic problem, the roads to the private subdivisions had to be opened up.[3]

People would speak then of the unreliable infrastructure with resignation and fatalism, with the Tagalog phrase *Bahala na*. Believed to come from the word *Bathalà*, the Tagalog name for God, the phrase is translated as "come what may" or "God will provide." People joke that in an archipelago perilously situated in the Ring of Fire and perpetually battered by volcanic eruptions, typhoons, and earthquakes, resignation is actually a reasonable if not wise attitude. It was also a useful response to missing infrastructure. "We have an important dinner, but we're stuck in traffic. *Bahala na.*" "The water was off all weekend, and there's not enough to bathe. *Bahala na.*" "Here in Makati, rush-hour traffic is bumper to bumper, but it's too polluted to walk. *Bahala na!*"

In fact, resignation was hardly the only force operating. By the mid-1990s, about half of the country's population was living in sizable towns and cities. As the administration of President Fidel Ramos pushed to turn the Philippines into an economic tiger, the environmental agenda shifted to the urban and industrial environments—the brown sector. Like the overuse of natural resources, the problems of the urban environment have deep political, economic, and cultural roots. Solving these problems requires major investments in public space, and it requires strengthening community programs; confronting inefficiency and corruption; finding ways for the government, private sector, and nongovernmental organizations (NGOs) to work together; and coordinating with other countries at the regional level. Because of their scale, potential solutions are expensive and complicated, and they require time and political will.

An example is the quality of air in Manila, which is among the most polluted in the world. Vehicles are responsible for the majority of the air pollution, for as much as an estimated 90 percent of the total suspended particles. The capital is home to more than one-third of all cars, buses, jeepneys, and motorcycles registered in the country. Between the early 1990s and 2003, the number of vehicles crowding Manila's streets increased from about 600,000 to 1.4 million. The congestion, poorly maintained engines, unruly driving, inferior quality of gasoline (only partially deleaded until 2001), and wide use of diesel all contribute to levels of air pollution that typically are far higher than the guidelines established by the World Health Organization.[4]

Medical researchers are only beginning to understand how air pollution affects human health. Recent studies show that long-term exposures to air pollution contribute to both illness (measured by missed days of work) and deaths that are considered premature. Air pollution is linked not only with specific respiratory ailments, such as asthma and chronic bronchitis, but also with high blood pressure, heart disease, and strokes. In the Philippines, 3 of the top recorded illnesses are related to respiratory disorders, as are 4 of the top 10 causes of death. Children are even more vulnerable to air pollution than adults, and nearly half of all infants' deaths in the country are caused by lung-related conditions.[5]

According to the national transportation department, it will take at least 15 years before enough roads and mass transit can be constructed to reduce the city's congestion. Even so, the airshed over Metro Manila affects seven different provinces with distinct governments, industries, and residents. To enforce the Clean Air Act, said a task manager at the Asian Development Bank, "we have to deal with mobile sources, stationary sources, cleaner fuels, cleaner technology. We have to monitor the air quality. We have to increase public awareness. We have to coordinate and monitor different agencies. We have to do transportation planning and traffic management, which is very complex. And most important, we have to do capacity building in agencies"—basically training people to do their jobs better. "But the environmental programs hardly have any staff. The country is full of unfunded laws, but even when they are funded, they don't have staff. And they don't have a budget either."

One of the first programs to tackle the challenges of the urban environment was the Metropolitan Environmental Improvement Program (MEIP), which was set up in a half dozen Asian cities by the World Bank and United Nations Development Fund. In the Philippines, MEIP was coordinated by Elisea (Bebet) Gozun, whose leadership of the program helped bring about her appointment as DENR secretary in 2002. Working with industry, local governments, and communities, MEIP laid the foundation for future programs. Its community initiatives included a wide range of pilot projects: to clean up rivers, recycle the waste from wholesale markets, create model recycling programs, teach communities to collect and manage their garbage. An anti-pollution campaign drew government agencies, industry, NGOs, and other groups to bar "smoke-belching" vehicles from the capital.

In targeting industry, MEIP had to recognize that few companies were complying with environmental regulations requiring them to report their emissions. Neither the government nor industry, Gozun said, had focused on source

Air pollution index, 1987. Manila.
MARISSA ROTH

reduction—limiting the release of pollutants—which is more effective and less expensive than cleaning up contaminants afterward. MEIP had identified Manila's top polluters, which included food processing plants, slaughterhouses and piggeries, beverage producers, dye and textile operations, and chemical factories. However, the government had inadequate enforcement power. "We're allergic to the word 'audit' here in the Philippines," quipped Gozun. To decrease emissions, the government encouraged firms to have a voluntary assessment done of their manufacturing processes that identified both ways to prevent or reduce solid waste and wastewater and also options for treatment. To make voluntary programs more effective, officials grounded them in deep cultural norms. Business leaders, for example, might respond to a campaign that publicly disclosed major polluters out of an aversion to *hiyâ*, shame.[6]

The project with the highest profile was the effort to clean up the Pasig River, once a main artery of commerce as well as a site of grandeur overlooked by Malacañang, the presidential palace. Cleaning the Pasig had been a strategic choice, and one with a broad appeal. While a river no longer slowed by garbage, sewage, and abandoned cars might not necessarily trigger economic development, it would offer clear evidence that anyone could see—and smell—that seemingly insurmountable urban environmental problems could be tackled. The Pasig also had great sentimental value, and it served as a symbol of the smaller, cleaner city that many residents still remembered. An earlier cleanup project from the Aquino era also tried to evoke memories of the city's waterways. Years later, its painted slogans remained on buildings and bridges: *Ilog ko, irog ko*, my river, my darling. Encouraged another, "Revive the River of Our Dreams."

Broad and serpentine, the Pasig River forms the trunk of a network of waterways that twist across much of Metropolitan Manila. The river cuts roughly east to west for 15 miles, crossing the narrow waist of land that rises between Manila Bay and Laguna de Bay. Just a few decades earlier, the Pasig, like provincial rivers, had been a place where people could fish and swim, wash dishes and clothes, bathe, and even gather drinking water. By the 1980s, the river had become the dumping place of last resort. Waste ranging from untreated sewage to industrial runoff all ended up in the Pasig. The river flowed sluggishly, the *esteros* and tributaries that fed it so stagnant that they seemed waterways in name only.

A joint effort of the Philippine and Danish governments, the Pasig River Rehabilitation Program (PRRP) was launched in 1993 with the goal of making the Pasig clean enough for boating and industrial use. Like other projects aimed at cleaning up the urban environment, PRRP faced major challenges. It had to curb industrial pollution, sewage, and solid waste; clean the riverways; and improve their banks. It also had to coordinate the governments of the nine riverside cities and municipalities and the many agencies whose responsibilities generally overlapped. To accomplish all this, the program also had to survive the transition from one presidency to the next. To begin, PRRP established a secretariat within the DENR to coordinate the cleanup and provide technical support. One of the more difficult early tasks was winning the support of the public, which saw the waterways as a place to dump waste. Said a representative of the secretariat, "We're trying to create an awareness that the river has been important in the development of Metro Manila and that everyone had a share in polluting it, so we all should have a share in cleaning it."

The program was controversial. Stretches of the Pasig had become the domain of the poor, the banks lined with rows of houses built out over the water on stilts. Although intended to decrease all the main sources of waste—domestic, commercial, and industrial—the cleanup focused on the squatters and ultimately became a way to clear their settlements as well as to clean the river. The course that PRRP took and the priorities it set show how difficult it can be to tackle urban environmental problems and how easily such projects can be diverted toward other goals.

◆

Manila's original name was Maynilad, or *may nilad*, which roughly means a place of *nilad*, a lily that flourished along the Pasig. In 1571, the string of Muslim villages clustered along Manila Bay and at the mouth of the Pasig were destroyed when the Spanish conqueror Miguel López de Legazpi defeated Raja Sulayman, the Muslim ruler who controlled trade in Maynilad. Legazpi renamed the place Manila, in Tagalog still called *Maynila*. There he built Intramuros, Spain's colonial seat in the region. Fortified by a moat as well as thick walls, Intramuros contained such impressive buildings as a cathedral, the governor's palace, and solid stone houses shielded by walls as high as 18 feet. Legazpi sought security, but he also wanted Intramuros to emanate power and inspire awe.[7]

When the United States took control of the Philippine Islands more than 300 years later, Manila was a city of about 250,000. The Spaniards, notorious for neglecting the economic well-being of their colonies, had done little to develop the city. Public health measures were being put in place in U.S. cities, and early annual reports sent by the colonial—insular—government to Washington decried Manila's condition. Year after year, they chronicled "municipal problems" that were "so full of difficulty that they need the best men that can be found to solve them." To make Manila a proper colonial city and integrate the country into the U.S. market, the American functionaries setting up the new government laid plans to build water and sewerage systems and other infrastructure.

Affordable housing was in short supply, and in 1904, an estimated 15,000 people lived on vessels along the Pasig and *esteros*. In a section titled "Overcrowding," the Philippine Commission's annual report lamented that "the city was started wrong, and it has remained wrong ever since." Manila had, it describes, three classes of "habitations": the modern, wooden houses, which were "sanitary and inviting"; the Spanish houses of "heavy masonry," which were "dark and poorly ventilated"; and the elevated *nipa* houses, for which, "unfortunately the requirements as to sanitation are not usually considered." The *nipa* districts, where landlords rented land by the square meter, also were overcrowded.[8]

When the U.S. architect Daniel Burnham arrived in 1904 to redesign Manila, he also saw the city as "wrong." Burnham spent about 40 days in Manila and, upon his return, wrote that it was an unplanned city "ill-suited for the abode of white men. The plan for the development of the city should make it, not only healthful but beautiful as well." Like others in that era's City Beautiful movement, Burnham believed in general plans for cities, in sweeping boulevards, in green parks and towering civic buildings, and in the idea that beautifying a city could make it more prosperous.[9] While many of his plans were not implemented for decades, he left his stamp on the capital, as he had on Washington, D.C.; my home town of Cleveland, Ohio; and Chicago, whose lakefront, studded with pale, imposing buildings, eerily resembles Manila's own waterfront.

After the extensive bombing at the end of World War II, huge portions of Manila needed to be rebuilt. It was only then, as the country industrialized, that the capital region began to expand rapidly, which it continued to do when Imelda Marcos governed the city. With her penchant for excess, she turned pockets of the city into showcases for cultural centers and luxury hotels, and despite the need, there was little investment toward housing the growing population. In time, many of the migrants from other provinces ended up living in illegal squatters' settlements, often on the thousands of acres of land left vacant by the leapfrogging development. In some sections of the metropolis, squatters made up about one-third to half of the population or more.[10] Looking from above like a patchwork of tiny, irregular roofs, their settlements spread over open fields beside government buildings in Quezon City, the largest residential section of Manila. Thousands of families occupied a stretch of reclaimed land between the waterfront and the elegant Roxas Boulevard. Their dwellings lined the Pasig River, and people even lived along the banks of the fetid *esteros*.

•   •   •

In her 1961 book, *The Death and Life of Great American Cities*, writer Jane Jacobs attacked the projects of urban "renewal" then being used to remake U.S. cities. "I like dense cities best and care about them the most," she wrote. Rather than addressing the economic reasons that bustling neighborhoods might turn into slums, officials, business leaders, and planners tried to eliminate and "renew" the blighted areas. Jacobs reserved particularly harsh words for the treatment of poor city residents. Those planning new projects, she wrote, expected that whole communities could be emptied, razed, and reconstructed, the people living in them relocated to new lives, "like grains of sand, or electrons or billiard balls. The larger the number of uprooted, the more easily they could be planned for on the basis of mathematical averages. On this basis it was actually intellectually easy and sane to contemplate clearance of all slums and re-sorting of people in ten years."[11]

In the United States, these urban renewal programs razed whole sections of cities, often to stimulate investment and rebuild decaying downtown areas. There was little respect for the lives being led there, the neighborhoods' deep roots, the institutions that held them together. Some poor communities were replaced with highways, some with commercial development, and some with

Living along the
Pasig River.
RYAN ANSON

impersonal housing projects that would foster more urban ills than they replaced. Other areas were cleared and not rebuilt; under the guise of urban renewal, slums were simply excised.[12] In a similar way, the demolition of squatters' settlements in Manila, particularly along the Pasig River, had more to do with the elimination of poor communities than with the role they might have played in fouling the river.

At the start of the rainy season in 1997, I joined a group of architects taking a ferry tour of the Pasig. Leading the tour was a woman from PRRP I will call Imelda. An enthusiastic advocate of the river and its future, Imelda was trying to interest the architects in redesigning the riverbanks. Ours would be a tour of the river and also the imagination, she said. The cleanup had made enough progress to make the ferry ride possible, but we still would be seeing the banks of the Pasig not as they were, but as they might one day be. For that, she warned, "you will have to use a lot of imagination."

We embarked at the walled medieval city of Intramuros, one of the few historic sites to have survived the World War II bombing of Manila. From there, the boat followed the twists of the Pasig past the neighborhoods of old Manila—the Chinese section in Binondo, the tangle of roads leading to the old church and market at Quiapo, then Paco and Pandacan. We passed the two tributaries that join the Pasig from the north: the Navotas River, which runs through the slum of Tondo and Caloocan City, and the highly polluted San Juan River, which flows through Quezon City before its waters, inky as a late evening shadow, enter the Pasig. We followed the river as it curved, then straightened out at the border between Mandaluyong and Makati, coming ashore just before we reached EDSA, famous for the protesters that thronged there in the days preceding the overthrow of Marcos.

We spent several hours on the water. The Pasig is shallow and in some places flows low beneath its banks. The cleanup had made noticeable progress. The unmistakable odor that once made nearing the river unpleasant had subsided, and we enjoyed ferrying past sights that most of us had never seen before. Plastic bags no longer clotted the water's surface, and dredging had removed so much debris that for the first time in years the water was navigable. Ruffled by a perceptible current, the wide, brown water looked and moved like a river again.

Although we passed only a few *bancas* and anchored oil tankers, boats had once been such an important mode of transport that firms, factories, and storage tanks lined the shores. The river revealed a grand side of the city; we floated past the National Archives, the National Press Club, the old post office. Bridges arced overhead: Jones, Quezon, Ayala, Nagtahan. We ferried past the San Miguel Building and the gold dome of Malacañang, whose large rooms, built to be open and cooled by river breezes, had been sealed against the water's stench.

The trip also revealed how much daily life had once revolved around the river and how for some, despite the tainted water, it still did. We passed clothes drying on fences, a young man in a blue *banca* eating a plate of rice, children swimming. Eventually we passed houses on stilts, their backs pale with sun-weathered wood. "Those are squatter families that still need to be relocated," Imelda said. "They don't understand that they pose a danger to the river, and that the river poses a danger to them."

Imelda had a clear vision of what the Pasig could become. The riverside near Intramuros would be transformed with a café and a festival market for artists. There would be parks and promenades, sweeping landscapes, and, of course, cleaner water. "We could bring back the charm and splendor of the Pasig River," said one of the architects. "I can really envision it," said another. "Like Paris."

The tour was selective; Imelda ignored most of the industrial buildings, and her enthusiasm waned when it came to the people living along the banks. Not only were their homes unsightly, but their waste was polluting the river. The squatters became her refrain, and she lamented their role in the Pasig's disrepair. "If companies would only maintain their backyards, it would be okay, because the squatters wouldn't go there." She drew our attention to an area planted with trees and closely trimmed grass. "No one would dare build a shanty there," she said. The San Miguel building was "historical, but it was all squatters in front before." She pointed to a park that also had been "all squatters before. Removing squatters is part of our beautification program. It is the priority of the government, the number one thing the government has done."

• • •

Despite an ongoing debate over the fate of the squatters' settlements, under PRRP, local governments had begun clearing the banks, community by community, tearing down the settlements and moving their inhabitants elsewhere. Thousands of homes had already been leveled. The largest remaining settlement stood in Mandaluyong along a section of the riverbank across from Makati, the financial capital, most modern part of the urban area then, and richest municipality in the country. This stretch of river was home to about 500 families. Barangay Barangka Ilaya, a neighborhood of about 85 families from different provinces, was next to be demolished.

Early one morning, I visited Barangka Ilaya, escorted by Klaid Sabangan, an organizer with the Manila-based Community Organization Training and Research Advocacy Institute (CO-TRAIN). The new constitution gave squatters some protection, requiring the government to compensate families being relocated and move them to habitable places with basic services. CO-TRAIN was one of a group of NGOs serving as advocates for the Pasig settlements to ensure that these conditions were met.

From the water, the houses along the Pasig had looked like rickety shacks cobbled together from scrap that perched over the river on stilts. When we ducked through the entrance to the longtime squatters' community, the narrow passageways flanked with wooden homes felt solid, like those of any crowded urban neighborhood. The community consisted of three long rows of houses running parallel to the river—those on the outer edge, the middle houses, and those on stilts. The narrow paths between them were hung with laundry, and festive bunting flapped overhead. Children darted across the paths, and we elbowed our way through them to speak with several women standing in their doorways. The stories that they told were stories about poverty in the provinces, corruption, and the expectations that people had of the government and its constitution.

Virginia (Vergie), a 49-year-old with four children, lived in a room furnished with a wooden table and bunk beds. A pair of gas burners and a set of aluminum pots served as the kitchen, and worn clothes hung near the ceiling. There was a television and a wooden cabinet with glass doors. "Eight people live here," said Vergie. "The four children, my mother, two grandchildren, a daughter-in-law—no, 10 people." The room measured about 12 by 15 feet.

Nearly 30 years earlier, Vergie had come to Manila from the eastern part of Samar, one of the more impoverished and typhoon-battered provinces. She and her husband had sold *copra*, the dried coconut from which oil is pressed. "When we got married we couldn't have a permanent home. We earned little money, so we wanted to go somewhere we could earn a living." She had graduated from high school, she said, and she wanted her children to have the same opportunity. "The schools were very far from our place; they had to travel by boat four hours."

Her story was not unusual. Her family had remained by the Pasig for 20 years, unable to afford to move. Their efforts to find better housing had also been derailed by corrupt officials. After their first home in Manila burned, the mayor had allowed them to move to the Pasig, but they also asked the landlord to build them apartments nearby. "When they were finished, the mayor's bodyguard occupied them. So we stayed where we were and built these shanties."

Some of the first families "demolished" along the Pasig were moved to a relocation site in Caloocan City known as Bagong Silang, which means "new born." Set up in the 1980s at what was then Manila's northern edge, it was unsanitary and the air was polluted. Although it was relatively close to the central part of the city, it lacked water, electricity, and dependable *hanapbuhay*—a way to learn a living. The next series of demolitions sent families about 40 miles south of Manila to Dasmariñas in the province of Cavite, where squatters had first been resettled 30 years earlier. Farther from the city than Caloocan, Dasmariñas offered even less access to jobs. By 1997, about a third of the residents relocated there had either sold or abandoned their homes and returned to Manila to work.[13]

The river's water lapped under Vergie's floorboards. A boat passed, and when the wake reached Barangay Barangka Ilaya, as must have happened many times a day, black water splashed up through the gaps. "There have been many problems," said Vergie. "Some of us were brought to Bagong Silang in Caloocan City. My mother was there, and one of my brothers. They were miserable. There was no water, no lights, no *hanapbuhay*, no schools, no basic facilities. Of course people complained, but nothing happened. I visited every week. We had to bring them water from here."

Vergie's neighbors' home had been demolished three years earlier, and they were brought to Dasmariñas "even though the site wasn't finished." As Vergie faced the threat of eviction, she "worried that the same thing might happen to us. We saw what happened with the previous relocation. About half of the families have left Bagong Silang and gone to places where they could find work. Some of them live under the Quiapo Bridge in Manila. The families are really affected. The children can't eat; they can't go to school. *Ayaw ko!* We don't want this.

"We asked the government to please do something for us, to develop the site so we wouldn't have to go far from our *hanapbuhay.* The government promised that there were many factories, many opportunities for jobs, but until now, *wala,* there's nothing. In December, President Ramos went and gave gifts—casseroles and thermoses. People said, 'Why do you give us these when we have no rice?'"

A claim used to justify the clearing of the squatters was that many were actually "professional squatters" who did not need the free or low-rent housing. When asked if there were such "professionals" in Barangka Ilaya, Vergie took offense. "They call us *magnanakaw* [robbers] and killers," she said. "But that is not true. My husband earns 200 pesos [seven or eight dollars] a day working at a construction site, but it's not enough. I used to work in a garment factory, but I got sick, and the factory was transferred to a very far place. We are trying to live a peaceful life, but we have very little education and we can't get better jobs."

It did not take long to pick up the terms that peppered talk about the Pasig. The houses built directly over the water were called homes, families, or simply people "on stilts." Before a "demolition," a census counted the inhabitants, and "structures" were "identified and tagged." The people themselves were "removed," with or without the building materials and possessions that had formed the roofs, walls, and contents of their homes.

One of the main objections to the demolitions was that residents were not given adequate warning. As in the sugar workers' villages in Negros Occidental, official notice of the looming eviction came as an anonymous sign. Tacked to Vergie's house was a white card, evidence of the recent "census-and-tagging operation" that had counted the families to be moved. The card read, "Mandaluyong, Baka-Ilaya area. #97-5-011." The last name of the "structure owner" was illegible, as was the tagger's name. When the census was done, the families did not know its purpose, said Vergie, so when the local government came to do the tagging, they were taken by surprise.

Although Vergie took offense at the suggestion that squatters might be "professionals," some of her neighbors, despite their deep roots along the Pasig, were better off than she knew. One woman lived in a small room with four of her five children. Her husband, she said, was "a long story"—a computer administrator who for nearly 20 years had worked in Riyadh, where he had another family. She and her second son, a college-educated bookkeeper in a bank, supported her family. "I was born in this neighborhood, and my mother was born here. If we are forced to move, we will move. My son bought a unit in a high-rise in Novaliches." She looked at us warily, realizing what she had let slip. "That's confidential," she said.

The women in Barangay Barangka Ilaya obviously knew that the waste from their houses drained into the river, but, caught up in the demands of daily life, they seemed unaware of what that might mean for their families' health—or that of others downstream. When Vergie received a ticket for littering, her main worry was the steep fine: "Four communities were issued violation tickets. We were fined 500 pesos [about 20 dollars]. We were concerned that we would get these tickets every day, just because we don't have a septic tank. We asked the mayor to help us. He told us to stay temporarily, and we didn't pay the fine. We built a septic tank—ourselves—with hollow blocks, cement, and sand."

· · ·

According to PRRP, when the cleanup of the river began, there were at least 12,800 families living along its banks, although an NGO counted some 60,000 families between Laguna de Bay and Manila Bay. About 2,300 families lived on stilts directly over the water. By the mid-1990s, thousands of those houses had been demolished. Squatters were being uprooted across the city, as in the 1960s and 1970s when huge urban renewal programs had resettled squatters from Manila. This time, the constitution gave them some protection; squatters being relocated were supposed to be either compensated or moved to a developed area where they could live. To ensure that these conditions were met, a coalition of NGOs and people's organizations working to "rehabilitate" the Pasig had formed the Sagip Pasig Movement—the Movement to Save the Pasig—which worked with riverside communities.

CO-TRAIN became involved early in this process. The NGO was started by activists with 20 years of experience in what its then executive director Corazon (Dinky) Juliano Soliman once called the "rough and tumble world of Philippine politics."[14] Its programs included workshops to improve the skills of community organizers, activists, and others doing social and economic development work. Its brochure of "guiding principles" termed this work "the humble contribution of some battle-scarred organizers to the future generation of development workers."[15] CO-TRAIN also worked directly on specific campaigns, such as the Sagip Pasig Movement.

As outspoken advocates of the Pasig communities, CO-TRAIN organizers helped negotiate the terms of the evictions and train organizers. To decrease solid waste dumped into the water, they set up recycling and garbage collection programs in the riverside settlements. "We're trying to organize the squatters' communities to be the equivalent of forest guards," said CO-TRAIN's interim executive director, Maria (Fides) Bagasao.

In Dasmariñas, CO-TRAIN helped those who had been relocated to settle in. Organizer Klaid Sabangan had worked with a number of communities along the Pasig and in Dasmariñas. "The demolitions even seem to have become more violent," he reflected. "If people don't move their belongings, they're confiscated. At one demolition, people were just moved in a dump truck and dumped outside Metro Manila."

In doing its work, CO-TRAIN often clashed with government officials and policy makers. "The local governments are not happy with us. We remind them about the importance of taking care of the citizens before removing them," said Bagasao. "Our organization's objective was to bring the families on stilts into the picture. We delayed the evictions until some provisions for housing loans could be made. We're not saying that they shouldn't be moved, but rather that the process should be sensitive and people should be compensated." She also questioned PRRP's priorities: "Will the river be an exclusive place for the rich, or will it be a public place for everyone?"

As it turned out, the squatters were not a main source of the Pasig's pollution. Tests of the river's water conducted in 1990 attributed 45 percent of the pollution to industrial waste and 45 percent to liquid domestic sewage. The remaining 10 percent, including the multitude of plastic bags that clogged the smaller waterways and triggered flooding, was considered domestic solid waste. Although that waste probably came from a variety of sources, said Fides Bagasao, it was blamed on the squatters.

Several years after the cleanup began, new analysis of the Pasig's water revealed that the squatters' contribution to the pollution had diminished further. While the total load of pollution had remained fairly constant, its composition had changed. Industrial pollutants had dropped from 45 percent to 35 percent of the total load, and solid waste from 10 percent to 5 percent. The new tests showed that about 60 percent of the pollutants came from Manila's largely untreated domestic sewage, an increase from 45 percent. As the city's population grew, so did the amount of domestic sewage reaching the river. Even though the waste from industry and squatters declined, the overall pollution load remained the same.

According to the PRRP secretariat, there were about 10,000 to 12,000 firms of all sizes along the river, or about as many factories as there were families. The polluters were divided into four groups, according to the level of waste that they produced. The list of "worst offenders" included industrial plants, residential high-rises, and malls. Among these major polluters, 314 were situated in the river basin and 140 along the banks.[16]

As was the case with other environmental rules, the government did not enforce its regulations for water treatment and pollution control, and the efforts to reduce industrial waste remained largely voluntary. A program coordinator with the PRRP secretariat explained, "We engaged the top 25 in a clean-river pact: we won't close them if they decrease the pollution load. Some of the first group built waste-treatment facilities, and all of the third group. The DENR is now targeting high-rises to put up treatment facilities. There are no real incentives for industry to participate, though."

During the earlier cleanup of the Marikina River, a major tributary of the Pasig that branches north a few miles from Barangka Ilaya, residents were moved within the city of Marikina. Some Pasig residents hoped that their local government would follow this example. There were large open fields nearby where housing could be built. "We are pushing for relocation in the city of Pasig," said one of the women in Barangay Ilaya. "The mayor promised that we'd be relocated there, but the cost of the unit is very high. People have to pay 22,000 pesos [about 880 dollars] in equity first and 411 pesos [16 dollars] monthly. The government will give each family 7,500 pesos [300 dollars] to help with the down payment."

"When I was five," she said, "I used to live along the Pasig, and I saw the preciousness of the river. We could swim, we could use the river for washing our clothes. When I was about 12, I wondered why the river became so polluted. There were people living on stilts before, but the water was clean. So I'm wondering why the government is always blaming us. There are factories, and the population is growing—people are moving from the provinces to Manila. They cause the pollution. I don't believe that it's the people living along the river who contribute the most. The factories can move to nearby provinces or dispose of their waste better. Why not move them? They're earning millions. The communities are the most visible, so we get the most blame."

"The government focused on organics because it had the historical records and the laboratory capacity to test for them," said Gozun. Even after it was known that they were not a major contributor to the Pasig's waste, the squatters continued to be relocated—and in a way that circumvented the protection given them by the constitution. Some of the local governments, which bore responsibility for the demolitions, claimed that the houses over the water and on the banks' first few yards stood in a "danger zone" washed by the river's wake. Because of a provision in a housing law, they did not have to compensate or relocate people living in danger zones. "Of the thousands evicted this year, most were not relocated because of this provision," added Bagasao. The Danish government, which was funding PRRP, had recommended that to reduce waste from the communities, conditions there be improved with "on-site upgrading" and programs to collect garbage and build septic tanks. "But the first move of the government was to move the people on stilts," said Bagasao, who described the policy as "based on scapegoating. In the name of the environment, they remove the poor people."

⚜

At first glance, the resettlement site in Dasmariñas, Cavite, seemed a grim alternative to squatting. There were no cities or towns, no shopping malls, no significant source of *hanapbuhay*, no bus terminals or other transport that might support a community of thousands. Called Bagong Bayan—meaning "new people"—the resettlement site covered more than 100 blocks, each bearing rows of one-story, square concrete houses with metal roofs. A lot was about 50 square meters, and residents had to pay 50 pesos a month—then about 2 dollars—for 25 years to gain full ownership. Although the place had the feeling of sameness, its houses varied. Some were carefully constructed; others were ramshackle. Some plain homes had extensive vegetable gardens with staked vines and young fruit trees. About 3,800 families lived on the site, 2,800 of them from the Pasig.

Although Klaid worked to prevent people from being moved to Bagong Bayan, he pointed out that the children we saw—and there were fewer of them in the streets than in Barangka Ilaya—looked healthier: "Before, they had skin diseases and their hair was thin. They looked sick. Now they look well."

Nonetheless, he recalled that when the first 300 families arrived, the transition was difficult: "The communities along the Pasig were clustered, and each cluster had its own character. Some were calm and organized, and the people cared for each other. Some were unorganized; the people were unruly, and they gambled. At first I couldn't imagine what would happen with all these people mixed together. In the beginning there was so much chaos. Even now, you can still see the worry in their eyes. At the start people also were clannish; they would introduce themselves by telling where they had been from. Now they're just from Cavite."

Although Bagong Bayan was supposed to be developed, like the Caloocan site it lacked schools and adequate utilities. Garbage and concrete rubble and other construction waste lay piled on some streets, and trash clogged stretches of the narrow drainage ditches lining the roads. The fluid the ditches channeled, which was clouded with the long, white filamentous bacteria that can indicate the presence of raw sewage, had flowed out and pooled in the streets.

"There were lots and poles for electricity and two water tanks, but no water," said Klaid. "The water supply is private, and you need to pay before being connected. It costs 1,800 pesos [more than 70 dollars] to be connected; only about

10 to 15 percent of the families have connections. The rest buy from them or just take water from the pipes, usually at night." Despite the refuse and sewage, the community was still cleaner and less congested than the riverside sites. Even with the drawbacks of the site, the unmet promises, and the shock of the demolition and move, not everyone regretted the change.

"It wasn't a good place to live," said another woman, "so it's only right that we should be moved. But if the government constructed a dike to control the river, it would be a good place to live, too. If the government thought of other things like that, we could have built our homes on top of the dike. We could have controlled the solid waste. So the government didn't necessarily have to move us to a faraway place. People were willing to pay for the materials and construction of the dike, but it wasn't an option."

Lorna lived on Block 73 with her husband and four children, ages 16 to 7 years. She was 35, her husband 45. Their main room contained a wooden sofa, two matching wooden chairs, and a cabinet holding books, papers, and half a skein of blue yarn. Bowling trophies filled a corner of the room below an elementary school certificate and two paintings of Mother Mary. The floor was concrete, not open planks as in the houses on stilts. Several rooms were devoted to a store and a bakery, and on the threshold, someone, probably Lorna, had pressed eight coins in a welcoming design.

Lorna and her husband had rented in several places before buying rights to a house in Santa Mesa along the San Juan River, which carries raw sewage from Quezon City to the Pasig. "We didn't live on stilts, but in the middle," she said. "The water [of the San Juan] is black and sticky, and it bubbles, especially at noontime. During high tide, it would come in the house."

Lorna's family was relocated in 1995. "We'd been renting in Mandaluyong for 3,000 pesos [about 120 dollars] a month. He [her husband] saved money from his salary—he earned 3,000 pesos a month—and bought the rights [to the house]. I wasn't there. I had to go back to the province. When I returned, we already had the house. He didn't ask; he just told me. It was about 20 square meters up and down, 40 square meters total, less than half the size of this [house]. It had four rooms. I rented out three for 500 pesos a month."

The house along the river had brought her a measure of security, despite its shortcomings. "It was dirty there," she remembered. "Every day there was fighting—between neighbors, between men and women, between men and men. Usually the community doesn't sleep. There was gambling, and people would drink all night. So our store was open 24 hours. Life was not that hard. We have a bakery here, but sales were much better there. People were working and had plenty of money. If you're going to open a store in the area, it's a good business

because people are lazy. They buy everything. They even buy their coffee ready-made so they don't have to make it and wash the cup.

"My husband worked in a slaughterhouse. Now he's a jeepney driver [in Manila]. He earns 200 pesos a day and works three times a week. He alternates; no one could stand the traffic every day. I'm the secretary of our day-care center and store. I buy the ingredients for the bread at the market and supervise the bakery. I sleep at 11 o'clock. We sell hot *pandisál* [salt bread] starting at five in the morning. There's less fighting here, and we live in a nicer place, but *hanapbuhay* is harder so life is hard."

Lorna acknowledged that some residents took advantage of the inexpensive housing in Bagong Bayan. "This is a place of no opportunity, so they sold. Some never even arrived here; their feet never touched the lots. Two families here sold their lots two months ago for 35,000 pesos [about 1,400 dollars] each, but they don't have money anymore. They paid off their loans, engaged in some vices. They had a good time. They're renting here now. Others went back to Manila to squatters' areas, where they'll meet demolition teams again.

"We plan to stay. It's hard for us to move and make a house again, start from scratch. And it's hard to socialize with new neighbors. We're familiar with the area. Everyone knows us; there's a feeling of community now. I want my kids to grow up here. And there's nowhere else to go."

The greatest challenge that the relocated families faced was the lack of work. In Barangka Ilaya, Vergie had told me about the model livelihood project set up by CO-TRAIN. "There's no *hanapbuhay* at Dasmariñas, only making rugs," she said. Lorna participated in the CO-TRAIN project, as did 43-year-old Carmen, who lived with her three children on Block 50 in a house with a yard of fruit trees. As we talked outside, Carmen picked up pieces of blue scrap cloth from a pile, cut them into narrow lengths, knotted their ends together, and wound them onto a ball the size of a *pomelo*, a fruit similar to a grapefruit. One of her daughters quickly looped the fabric on a loom to make a rug.

Carmen described the project's financial nuts and bolts, repeating the numbers that had worried Vergie: "Doormats sell for six and a half to seven pesos. Maybe 15 pesos in Manila. We pay 15 pesos per kilo for cloth scraps from an assembly factory. We can make four mats from one kilo of scraps. As a family, we can make 10 to 15 doormats a day. Today we bought four kilos. The whole family worked. Tomorrow, 16 doormats will be ready. We don't do it every day. Sometimes we do it for a hobby, after we watch TV or run around."

Not including the cost of the fabric, the 16 mats would bring them 65 to 180 pesos—3 to 7 dollars—depending on where they were sold. With the higher price they brought in Manila, the mats would almost cover the rent, but not other household expenses. A small bag of mangoes or a whole chicken from the wet market cost about 100 pesos. Nevertheless, a man working at the pier might earn only 185 pesos in a day and also have to pay 30 pesos to travel each way. In Dasmariñas, people lived close to the margin, and every peso counted.

．．．

Many explanations of why the Philippines failed to thrive economically after the 1960s point to the dynamics between the government and the local elites. Widespread corruption and the power of the elites, such as the sugar families in Negros, hobbled the government's ability to stimulate the economy. "[T]he central government can only dispense largesse; it cannot finance economic growth," wrote one analyst in the mid-1990s.[17] According to Arsenio Balisacan, an economist at the University of the Philippines who has twice served as an undersecretary in the Department of Agrarian Reform, national economic policies in the Philippines have not promoted job growth, particularly in rural areas. The outcome, as in other developing countries, has been the mushrooming of work considered separate from the economy's more formal sectors, such as manufacturing, service industries, and retail. In the mid-1990s, Balisacan estimated that, outside of agriculture, the informal economy employed about 80 percent of the workforce.[18]

Government policies also encourage Filipinos to find work overseas, as they have for about a century. In the early 1900s, a wave of workers labored in Hawaii. Others fanned out around the world after World War II. Rather than creating jobs at home, Ferdinand Marcos launched a program to promote working abroad, and subsequent administrations have increasingly used overseas work to ease the country's perpetually high unemployment. Toward the end of the 20th century, Filipino workers provided needed labor as countries in the Middle East industrialized. During the second Gulf War, more than 4,000 worked in Iraq, mostly at military installations. These workers typically are educated men and women who are supporting their families, often by working as maids or laborers. In 2004, the government reported more than 1 million Filipinos working abroad and sending home "remittances" estimated at nearly 70 billion pesos (about 1.3 billion dollars). These official figures fall far below those I gathered in the early 1990s, which counted about 3.5 million Filipinos working abroad legally and 1.5 million illegally, even though the number of overseas workers has climbed since then. Another source, which omitted illegally sent remittances, cited more than 7 million overseas workers who in 2003 channeled 7.6 billion dollars through Philippine banks.[19]

The effects of these government policies reverberate in daily life, from the poorest provinces to the banks of the Pasig. Carmen was among those who worked in the uncounted informal sector. "My main source of income is laundry. When we were in Manila, I earned more because I could bring the clothes to my home. I could go to dormitories and find customers everywhere. They paid three pesos for a shirt, four pesos for pants. Here, I can't bring the laundry back to Dasmariñas, so I do it there, and it limits the number of people I can work for." Years earlier, the trip into Manila from adjacent provinces had been relatively quick, but as new housing and commercial developments sprawled over those intervening miles, the narrow roads filled with traffic. To keep their commute to about an hour, Carmen and her husband would leave at four o'clock or four-thirty in the morning and return at eleven at night. "I go to Manila sometimes twice a week, sometimes every day," Carmen said. "I earn 300 pesos now. Before, I would earn 400 pesos minimum."

Like millions of others, Carmen had also worked overseas. When her husband refused to pay for a new home, she was determined to "provide." In four months as a maid in Hong Kong, she earned enough to buy a house on stilts near Manila Bay. They lived there for nine months. "We transferred there on June 14, before the new house was finished," she said. "The census started July 14." They moved to Dasmariñas in March.

Filling water jugs. Malate, Manila.
MARISSA ROTH

"We didn't agree to be demolished," Carmen pointed out. "The demolition crew of city hall did the dismantling. I guarded them while they were dismantling my house, because if you guard them they'll do it slowly and with care. So I didn't have the experience that the others did. There was a feeling of disappointment, of course, to see my house dismantled. The crew had a hard time. They told me it was durable. Of course it was durable. It was new."

Carmen still had mixed feelings about the move. As we talked, she continued to work, her hands automatically cutting, knotting, and winding. "It's right for us to be relocated in a good place like this. Although it's right to earn a living in Manila, the environment is difficult—there's violence, different kinds of vices. It's *magulo* [unruly]. I like it here now. The place is clean and it doesn't have a foul odor. I prefer to stay here, but work is still in Manila. If I sell my lot, that means I will rent in Manila. That will be harder. It was better there. *Malapit sa trabaho, malapit sa pera* [work was near, money was near]. That's why people sold their lots and returned."

CO-TRAIN workers could not prevent the squatters' settlements from being relocated, but they could delay the move, and they could try to improve the demolitions and the conditions of the resettlement sites. While providing work was not their main focus, they set up the rug-making project as a model. As organizers have learned in fishing and upland villages, although such projects are needed, it is hard to make them succeed. Local people may find the new work too unfamiliar, it may be difficult to create links with local markets, and the income may be inadequate or unreliable. As often happens with environmental projects set up in rural communities, the sudden infusion of funding—whether from loans or grants—also can backfire.

When we left Bagong Bayan, Klaid told me that Lorna and Carmen, who had been leaders of the rug-making project, had omitted several key facts. "People here don't want collective projects. They're very individualistic," he said generously. They also were not always honest. The rug project, begun with a small loan from the DENR, was run by the community organization, which bought scrap cloth, gave it to the members, and paid them per mat for their labor. "Someone would take the mats to Manila to sell them," Klaid said. "The women who sold the mats never gave the money back, though, so the initial loan from the DENR was never repaid. It was about 12,000 pesos [about 480 dollars] or more." The women we talked with were among those who failed to repay the money, said Klaid, adding, "They needed it to feed their children."

❧

Like the Pasig River, Manila Bay is a symbol of both beauty and neglect. A deep inland bay, it is rimmed by Roxas Boulevard, which bears a yacht club, gracious hotels, the U.S. embassy, and grand civic buildings. The city faces away from the soft blue horizon. Until the early 2000s, when a restaurant-lined promenade was built along Roxas, it was possible to live in Manila and rarely see the bay, where much of the city's wastewater has been dumped.

When environmental lawyer Antonio Oposa spent a year at the Harvard Law School, he learned that a lawsuit had sparked the cleanup of the harbor in

Boston. He decided to try a similar tactic in Manila. In a speech delivered in 1999 at the elegant Manila Hotel near the bay, Oposa shocked the Manila Rotary Club with a little-known fact: most of the capital lacked sewerage. Striding back and forth, he challenged the businessmen, politicians, and diplomats sipping after-lunch coffee to try to solve a problem that affected them all. He began with the name of the government agency regulating Manila's water and sanitation systems, the Metropolitan Waterworks and Sewerage Services (MWSS). "The 'S' in MWSS stands for sewerage, and every month you pay a sewerage fee," he said, aiming for shock value, "but MWSS has been in operation 30 years and hasn't done anything to address sewage." The well-heeled audience squirmed. "There's almost no sewage treatment and no toxic and hazardous waste facility. Five million gallons of raw sewage go into Manila Bay every day. We need to stop thinking of Manila Bay as our toilet."[20]

In 1993, Oposa had successfully sued the government on behalf of a group of minors, including his young children, whom the case describes as deprived of their constitutional right to enjoy a "balanced and healthful ecology in accord with the rhythm and harmony of nature."[21] On behalf of citizens living near Manila Bay, which is so contaminated that it poses a public health hazard, he filed a class action lawsuit against a long list of corporations and government agencies which, he claimed, bore joint responsibility for the bay's pollution and should be responsible for its cleanup. In 2002, the court ruled in his favor.

The case was one of a series of steps being taken to improve the city's water and sewerage systems. While the Pasig cleanup was among the more visible environmental projects, improving the water and wastewater systems would bring far more profound changes to daily life. As in other megacities, cleaning the urban environment meant imposing necessary infrastructure on an otherwise intact metropolis—a technically challenging task that is expensive and politically daunting. Manila's outer face was already changing quickly as highways, flyovers (overpasses), underpasses, elevated light-rail lines, and medium-rise buildings popped up across the skyline. Improving the water and sewerage would require even more disruption and construction. It would also require a remake of MWSS. Notorious for inefficiency and corruption, the agency was said to be so "overstaffed" that it was jokingly called "more of an employment agency than a water authority."[22] To overhaul MWSS, the administration of Fidel Ramos contracted out the agency's services to private industry.

Most of Metro Manila's water traces to the Angat River, which flows south from the western Sierra Madre mountains into the huge Angat reservoir in Bulacan before being diverted to treatment plants closer to the city. In 2001, a long-overdue 15-mile tunnel to the Umiray River in Quezon province was finally completed, boosting the water supply further. While poor watershed management and deforestation have helped degrade the water supplies elsewhere in the Philippines, in Manila the main problem has been water delivery.

In the 1990s, half to two-thirds of Manila's residents were connected to the public water system, including about 20 to 25 percent of poor households. Those without access to running water might buy water privately from hawkers, who jacked up the price as much as 10 to 20 times. Those with water connections experienced low water pressure and unreliable service, and rationing left much of the city without running water for large parts of the day. Even in large office and government buildings, to flush toilets and wash hands, people scooped water from large, covered containers. Homes in affluent neighborhoods could have running water 12 or 14 hours a day while elsewhere the water might be "on" far less than that. In some areas, including a place where I lived one summer, the water would run for a single hour at daybreak, when maids would run to fill all the containers. Those who could boiled their drinking water.

Urban residents always know the deficiencies of their water system, but only infrastructure mavens contemplate where sewage pipes lead. Manila's main sewerage system, built under the Americans in the early 1900s, serves about 500,000 people. Makati, whose affluent villages provide their own security, garbage collection, and traffic control, also has its own waste treatment plant. Other homes typically rely on septic tanks—Gozun estimated about 650,000 in the capital—which generally have not been pumped properly or regularly. Even in 2000, the water companies reported cleaning only about 1,600 tanks. Laws requiring malls, factories, and high-rises to build and operate wastewater treatment facilities were rarely enforced, and despite the gains of the Pasig cleanup, much of the sewage went directly into the waterways or Manila Bay. While the stench of the *esteros* strongly suggested the presence of raw sewage, the public remained unaware of the status of what is politely called liquid domestic waste.

The 1995 Water Crisis Act gave President Fidel Ramos the power to privatize the beleaguered MWSS. At the recommendation of the International Finance Corporation, the investment arm of the World Bank, the city was divided into two zones, East and West. The rationale was that competition between two companies would bring better operations and more security; if one company failed, the other could take over. Privatization was expected to expand coverage in dry areas and bring better water delivery, lower costs, a 24-hour water supply, and, eventually, sewerage and sanitation. The 25-year contracts included a number of key provisions. They would guarantee the companies profits and set up a system of tariffs—fees—in which higher commercial rates for water subsidized lower residential ones. While allowing for price adjustments to compensate for significant foreign exchange losses, the contracts assumed that the Philippine peso would undergo only minor depreciation against the dollar. The firms would need to get approval from MWSS for all tariff increases, although none was expected for 10 years.

In 1997, the operations of MWSS were turned over to the lowest bidders. The Philippine constitution limits foreign ownership of businesses, so each new water company was jointly owned by an elite family and an international water company. Manila Water Company, Inc., took over the East Zone, with about 40 percent of the population, including the eastern suburbs and Makati. Manila Water was owned by the corporation of the Ayala family, descendants of the Spanish elite whose empire includes a 12-acre complex in Makati; the United Kingdom's United Utilities Ltd.; and the U.S. Bechtel Corporation. The corporation promised a water rate of 2.32 pesos per cubic meter.

Maynilad Water Services, Inc., took over the more densely populated West Zone, which contains about 60 percent of the population, including Caloocan and other northern suburbs, old Manila, parts of Quezon City, and the southern province of Cavite. Maynilad was led by Benpres Holdings, the corporation of the Lopez family, which built its wealth with the sugar industry and expanded into shipping, power, and the media. Its international partner was French Suez Lyonnaise des Eaux (now Ondeo Services). Established in the mid-1800s to build and operate the Suez Canal, Suez was one of the largest private water corporations, with operations in about 130 countries. Maynilad promised a water rate of 4.97 pesos per cubic meter of water.

The water companies quickly ran into problems. The contract coincided with the 1997 financial crisis in East Asia which—even as they began to streamline their operations, lay water pipes, and uncover illegal connections—made it difficult for them to secure bank loans to finance the maintenance and improvements. The peso rapidly lost value, which hiked their foreign debt payments. Because the East Zone was smaller but required more investment for both water and sewerage, Manila Water had taken over only about 10 percent of the total debt from MWSS. Maynilad, with its larger customer base and higher rate, shouldered about 800 million dollars. As the value of the peso slipped from about 26 per U.S. dollar to more than 50, Maynilad took in revenues in pesos, but it had to repay the loans in dollars—which effectively doubled its debt.

The year 1997 also brought the El Niño drought, which reduced the water supply by about one-third. "We have always thought of ourselves as a water-rich country," said a lawyer working for Manila Water. "People see the rain and think that we have plenty of water, but water coming from the taps is something else. [During El Niño,] the volume in the dam was dropping every day." Maynilad Water, which already had lost expected revenues due to the delays in the Umiray tunneling project, was again hard hit.

Finally, the water companies faced greater problems than expected with illegal connections which, together with the leaks that plague any older water system, drained a staggering 60 percent of the total water pumped into the network. Like the fouling of the Pasig River, these "nonrevenue losses" were often blamed on squatters, who might illegally attach a shared standpipe to the central network to bring running water to their settlements. Although squatters accounted for the largest number of illegal connections, the major volume of water siphoned illegally traced to larger perpetrators. Speaking on condition of anonymity, a high-level Maynilad Water official told me frankly in an interview, "The squatters' areas are not a major part of the loss. The 60 percent of the water that is not paid for mainly goes to leaks and commercial losses."

While the official would not name names, he implied that some of those stealing water were well-known malls, companies, and apartment complexes, and he said the water company was changing the meters for the large customers and going through their accounts one by one, looking to see "whose rates seem too low." They also used metal detectors to look for pipes or measured the flow of water near buildings and downstream. The Philippine Center for Investigative Journalism was less discreet; it reported that large well-known customers harboring illegal connections included San Juan Slaughterhouse, Unilever Philippines, a housing developer, and the Coca-Cola bottling company.[23]

Under MWSS, the employees were integral to the illegal connections. They might take bribes to ignore the connections, look the other way as pipes were laid, or connect the pipes themselves. Some employees even sold water in dry communities, a lucrative sideline. After privatization, a small core of MWSS had remained as a regulatory body, but other employees went to work for the new water companies. With their workforce coming almost entirely from the old MWSS, the water companies could not readily change the culture of corruption. Companies and individuals used to pay off MWSS to connect to its network illegally, said the Maynilad Water official. "It still happens."

After less than a year, both water companies applied to raise their rates. In 2001, the concession agreement was amended to allow additional increases, and a series of hikes eventually more than quadrupled rates in both concessions. Nonetheless, while Manila Water became profitable within a few years, Maynilad Water did not. Maynilad carried the greater debt burden, but it also had far greater operating expenses; as it spent nearly twice as much to "produce" its water, it incurred growing losses. In 2003, after MWSS rejected yet another proposed rate increase, Maynilad attempted to withdraw from its contract and return the concession to the government. It also announced its intent to

Metro Manila.
RYAN ANSON

seek damages from MWSS, which it claimed had failed to honor clauses in the contract. An international arbitration court ruled that there were no grounds to end the contract, leaving MWSS and Maynilad to try to hammer out a compromise deal that would allow the Lopez family—Benpres—to leave the concession while ensuring uninterrupted water service for that part of the city.

When the contracts were signed, the privatization of MWSS was the largest such privatization of a water system in the world, and it was looked to as a harbinger of what other countries might expect. In some ways, it has been. At the start of the 21st century, an estimated 1 billion people worldwide still lack clean water, and 2 to 3 billion do not have access to sanitation. About 5 million, many of them infants and children, die each year from waterborne diseases. The growing need for fresh water has coincided with the new wave of globalization, creating a market for international firms to build and run systems to treat and distribute water.[24] In the 1980s and 1990s, countries from Latin America to Europe to Southeast Asia leased or sold their utilities to private firms, hoping for a safer, cheaper, and more reliable supply of water.

The question of whether or not private companies can run utilities better than governments remains. So far, in privatizations in a range of countries, parties on all sides have been disappointed—investors with their returns, governments with the systems' reliability, and consumers with their service and rates.[25] Proponents of privatization believe that private industry is better able to offer a reliable water supply, particularly in a country where the government is hobbled by corruption and inefficiency. Critics argue that water is a basic human right and that private control does not necessarily provide a safe, secure, and affordable supply and can even bring social costs. They also question whether vital public services should be entrusted to international companies that are driven by the profit motive and lack a long-term stake in a country.[26]

In Manila, there have been problems with both the private and public sectors. Manila Water continues to supply its customers with water, but at a much higher price than promised. The government-run MWSS itself failed to provide a sound and efficient water supply, and it left a legacy of corruption that privatization could not quickly undo. While it was the Filipino portion of Maynilad that tried to withdraw from the contract, its international partner has dissolved contracts in other countries. Analysts predict that Maynilad will be sold.

Privatizing MWSS did not fully benefit the large, poor population that had the least access to water. Even as the water companies expanded their service, poor communities remain disproportionately unserved. In both zones, many poor customers use communal standpipes, which can be run by private operators at elevated prices. They also buy high-priced water from vendors.[27]

Privatization did not resolve what Gozun called the biggest urban problem: wastewater. "People don't understand that we all contribute to it. It's never prioritized," she said in 1999. When Manila Water took over the East Zone, the wastewater of only 2 percent of the population was treated at a treatment plant, and nearly one-quarter of the population was not served by a sewerage system or septic tank at all. The company's own plans projected covering only 16 percent by 2006 and 51 percent by 2011.[28] As Metro Manila continued to grow, it would be hard for the construction of sewage facilities to keep pace. "The more water, the more sewage," said an official with the Pasig River secretariat. The water companies have said that they would "only address the sewage in five years. They should address it first."

. . .

On a recent trip to Manila, I reflected on how the capital has changed over the years since I first lived there. The population of the metropolis continues to grow, and wealth and poverty continue to coexist, even though some of the more visible squatters' settlements, including Barangka Ilaya, have been cleared. On fields once overgrown with grass have risen office buildings, light-festooned shopping malls, condominiums. The major infrastructure—once unreliable or inadequate—has begun to catch up with the city's growth. Despite the surge in vehicles, the traffic remarkably has eased a little. Highways have been completed, roads widened, and underpasses and overpasses constructed to eliminate long queues at major intersections. Modern light-rail lines march across major thoroughfares, reducing commutes from a grueling 1 or 2 hours to a predictable 20 minutes. Express bus routes and additional rail lines are being planned, and in congested neighborhoods, sidewalks and roads have been repaired and people walk once again.

While undeniably still polluted, the air is noticeably cleaner. Gasoline has been deleaded, and diesel—used by the buses and jeepneys that travel the most miles countrywide—contains 75 percent less sulfur. The majority of tricycles have switched from two-stroke to the less-polluting four-stroke engines. Fleets of buses run by compressed natural gas are being introduced, along with taxis fueled by liquefied petroleum gas. Although hampered by corruption, a system to inspect vehicles has been set up.

Despite the difficulties with the water companies and the rising cost, for many the availability of water has improved. In homes and offices, the once-indispensable dippers and standing containers of water have disappeared. Although expensive, power, too, has become more reliable; brown-outs seem a worry of the past. Sewerage remains minimal, but the water companies have divided their zones into districts to be served by small sewage treatment plants. Construction has begun on these plants, which will treat sewage and the contents of septic tanks.

Changes in policy have made it possible to anticipate future changes. In five years, three major environmental laws have been passed to clean up the air, water, and solid waste, providing the scaffolding to improve further the urban environment—and urban health. The worst-case vision of a chaotic urban nightmare has not materialized in Manila. In this huge, fast-paced city, incremental changes have added up to changes in the daily lives of many. It seems realistic to predict that they will continue to.

Epilogue

*I*n longstanding debates over the trade-offs necessary between ecological protection and economic growth, ecological degradation is often presented as an inevitable stage that a country must pass through as it develops. After all, the argument goes, countries that industrialized early expanded their manufacturing base without attempting to curb environmental and health effects—including those within workplaces—until later. Their approach separated industrialization from conservation and environmental protection, casting a clean environment as a luxury that only the wealthy could afford.

This version of development omits crucial details. As these countries developed, many of them did take measures to manage their natural resources, recognizing that depleted forests and fisheries or logged-over watersheds brought economic and ecological consequences that magnified each other. As their urban areas grew, citizens were safeguarded with running water, garbage collection, sewage systems, safer workplaces. The vital public health infrastructure improved people's health and enabled cities to flourish. Nonetheless, both the conservation and public health measures required considerable debate within their societies, as they still do today.

Also important in understanding conservation and environmental protection in industrialized countries is the fact that some of these measures came at the expense of other countries. After World War II, the United States and Japan were able to keep their forests well managed and their watersheds forested partly because they could rely on wood imported from Southeast Asia and elsewhere. Similarly, wealthy countries have been able to reduce their waste by diverting it: releasing it downstream, sending it downwind, dumping it in the ocean, or exporting it.[1] Even now, the United States exports huge quantities of used electronics, whose components can be hazardous when they are recycled or discarded.

Many countries developing today are caught between a reliance on their natural resources and insufficient economic alternatives. In the Philippines, the economy depends on the country's natural wealth. Half the population lives in the rural provinces, where agriculture and tourism continue to be important sectors. In addition, millions of people live at the margins, relying, usually at the subsistence level, on land and other natural resources. Until they have access to work and housing, they will continue to degrade the natural resources; the decline of those resources—from overuse as well as industrialization and urban growth—will continue to impoverish them and others. At the local level, where the erosion of the rural environment affects daily life, people understand this. "If industry comes in and we can't fish, how will we survive?" asked a council member in a coastal town in Mindanao. "We're not against industrialization, but we don't want to lose the ability to feed ourselves every day. We'd like our children to have jobs they can depend on. Maybe then coastal resources wouldn't be our primary concern."

Poorer countries also cannot necessarily export the by-products of industrialization to other countries, and with their rising populations and already unwieldy cities, they face unique challenges of cleaning up and managing industrial and domestic waste before they have finished building their urban infrastructure. Megacities such as Metro Manila are only now creating basic public health measures, especially in the informal settlements that house much of the population. It will be years before these measures are complete.

Although despoiling the environment may be portrayed as an unavoidable, temporary stage of economic development, many ecosystems, including those on tropical islands, are not resilient enough to recover from sustained degradation. There, rather than temporary, the changes can be irreversible. Rehabilitation—an odd word used for the remake of both criminals and damaged ecosystems—may be possible in some places, but in others it is not. Restoring ecosystems also frequently requires precise technical expertise—to model a fishery, dredge a lake, reconstruct a wetland—and prohibitive expense. The forests serve as a warning. Once a vital part of the economy, in 2003 they contributed only negligibly to the gross domestic product.[2]

Developing in a manner that can be sustained—including managing natural resources and tackling the frontier of the urban environment—is not a luxury. It is a sound development strategy. Acting otherwise brings costs—of replacing exhausted or depleted resources, cleaning up industrial waste, paying for the care and consequences of poor health, and, in a land-poor country, forgoing the use of ruined land. "We do sustainable development precisely so we preserve our options," said a former high-level official from the DENR. "Countries *can* be green tigers. It may be easier politically when the economy is growing, but we have to do it all the time. We can't afford not to."

Outside the archipelago, the Philippines is often remembered for the excesses of the Marcos era, the conflicts in its southernmost provinces, its ongoing political turmoil. It is also remembered for the chaotic state of its natural environment. Of more lasting importance are efforts to deal with that ecological decline. I have found that in this highly politicized society, people working on environmental issues—whether through the government, non-governmental organizations (NGOs), or religious and business sectors—tend not to be swayed by apparently easy answers. They understand all too well the deep social, economic, and political roots of conflicts over natural resources. They know the importance of devising mechanisms for local control, and even though their endeavors might not always succeed, they are developing ways to engage the different parts of society in addressing the costs of environmental decline. Despite recurring setbacks and the fatalism in the culture—the attitude of *bahala na*—they also tend to be memorably determined, with strong national pride and a sense of being part of a long history of efforts to gain independence and build their own country.

This is not to deny the difficulties of the current economic and political situations or the profound ecological damage that the country has sustained. The overharvesting has not stopped. Overfishing continues, as does illegal logging, and rural communities continue to suffer the consequences of both. In 2003 and 2004, mudslides in Leyte's mountains—not far from the site of the Ormoc tragedy—and in those of Quezon and Aurora provinces in Luzon killed hundreds of people. Still, since the transition to democracy brought an opening in which people could voice their concerns about the ecological degradation, the decline of the forests and coasts has diminished and the environmental movement has grown.

For those living with mudslides, shrinking catches of fish, and trying urban surroundings, the accomplishments of this environmental movement may be accruing slowly, but they are not trivial. They include strings of marine no-take zones, reforested watersheds, an expanded network of protected areas that covers most major ecological areas. Like the coral sanctuaries that reflect the approach to coastal conservation honed on Apo Island, these local projects tend to be grounded in communities while also linked with institutions at regional and national levels. In Manila, tangible progress has been made to reduce traffic, manage waste, provide safe water, make the air more breathable.

Many of the projects have been supported by foreign funds and some have faltered over the long term, but they offer innovative and important models that can be applied elsewhere. At the policy level, major environmental laws are in place, and the slow work of implementing them has begun.

In relatively few years, the environment has become a larger part of public debate. Although journalists faced death threats for reporting on logging in the late 1980s and early 1990s, environmental news has become standard fare, and people have better access to information, especially in urban areas. In cities and villages alike, officials have begun to recognize that a degraded environment brings costs and that conserving resources brings benefits. They also realize that the constituency for environmental issues has grown and that citizens vote for local officials who recognize those costs and benefits and understand whom they affect. The relationship between the government and NGOs is better established and less contentious. Some of the larger NGOs have evolved into stable institutions integrated into political life. The organizations may still serve a watchdog role, but it is one that grows less adversarial. In their joint efforts, NGO and government officials are now joined by representatives from the private sector as well.

Establishing and strengthening democratic institutions is a slow process, far slower than my own government has sometimes been willing to recognize. Rather than unfolding in a straightforward fashion, it also can be convoluted. Gains from one administration might be lost when that era ends, but they are not necessarily lost forever. The goal is to create a measure of continuity across administrations and ensure that, once begun, work headed in a constructive direction will endure. Environmental work can be part of this process.

Environmental projects such as those described in these pages do more than replant a logged-over forest, revive a degraded reef fishery, temper the acridity of Manila's air. They leave behind a foundation for future projects and future development. The education and organizing in communities, the technical training, the strengthening of organizations and institutions, the relationships among the government and businesses and NGOs, the tussles within the government, and the new policies also contribute to the long, slow building of a democracy. In these and many other ways, environmental work can serve to bolster communities, governments, and the economy. It does not oppose development; it is an important part of it.

Notes

Preface

1. With thanks to Lawrence H. Fuchs, *"Those Peculiar Americans": The Peace Corps and the American National Character* (New York: Meredith Press, 1967), p. 94.

Chapter 1

1. James K. Boyce, *The Philippines: The Political Economy of Growth and Impoverishment in the Marcos Era* (Honolulu: University of Hawaii Press and OECD Development Centre, 1993), p. 4.

2. Lawrence R. Heaney and Jacinto C. Regalado, Jr., *Vanishing Treasures of the Philippine Rain Forest* (Chicago: The Field Museum, 1998).

3. Republic of the Philippines, *1994 Philippine Yearbook* (Manila: National Statistics Office, 1995), p. 134.

Chapter 2

1. Some details of the Ormoc account have been drawn from Marites Dañguilan Vitug, *Power from the Forest: The Politics of Logging* (Manila: Philippine Center for Investigative Journalism, 1993), pp. 1–9; Hernani P. De Leon, "Conversion of Forest Lands to Agri Use Caused Floods," *BusinessWorld*, November 12, 1991; Hernani P. De Leon, "Leyte Disaster Blamed on 'Excessive Rainfall,'" *BusinessWorld*, November 11, 1991; Fel V. Maragay, "Factoran: I Will Name Leyte's Illegal Loggers," *Manila Standard*, November 11, 1991, p. 2; Santos Patinio, "Loss of Forest Cover Blamed," *Philippine Newsday*, November 11, 1991, p. 1; "Flood Exposes Wider Philippine Disaster," *Asian Wall Street Journal*, November 27, 1991.

2. Alfred Russel Wallace, "Physical Geography," in *The Malay Archipelago*, vol. 1. Available at www.worldwideschool.org.

3. Charles Darwin, *The Voyage of the Beagle* (New York: Bantam Books, 1958), entry for February 29, p. 10.

4. Alfred Russel Wallace, "Equatorial Vegetation," in *Tropical Nature and Other Essays* (1878), p. 30. Available at www.wku.edu/~smithch/wallace/S289.htm.

5. Wallace, "Equatorial Vegetation," p. 30.

6. Wallace, "Equatorial Vegetation," p. 67.

7. Quoted in Vitug, *Power from the Forest*, p. 11.

8. José Rizal, *The Lost Eden (Noli Me Tangere)*, trans. León Ma. Guerrero (New York: Norton, 1961), p. 53.

9. Lawrence R. Heaney and Jacinto C. Regalado, Jr., *Vanishing Treasures of the Philippine Rain Forest* (Chicago: The Field Museum, 1998), pp. 16–22.

10. Catherine Caufield, *In the Rainforest: Report from a Strange, Beautiful, Imperiled World* (Chicago: University of Chicago Press, 1991), pp. 65–76.

11. Quoted in Daniel J. Hillel, *Out of the Earth: Civilization and the Life of the Soil* (New York: Free Press, 1991), p. 104.

12. Robin Broad with John Cavanagh, *Plundering Paradise: The Struggle for the Environment in the Philippines* (Berkeley: University of California Press, 1993), pp. 56–72.

13. Jun Jabla, *Defending the Forest: A Case Study of San Fernando, Bukidnon, Philippines* (Davao, Mindanao, Philippines: Kinaiyahan Foundation, Inc., 1990), p. 18.

14. Jabla, *Defending the Forest*, pp. 30–31; Vitug, *Power from the Forest*.

15. Broad, *Plundering Paradise*, pp. 67–68.

16. Jabla, *Defending the Forest*, p. 30.

17. Ruth M. Esquillo, "Community Action on Forest Protection: The Case of San Fernando, Bukidnon" (master's thesis, Ateneo de Manila University, 1992), pp. 85–88.

18. Esquillo, "Community Action on Forest Protection," pp. 12–13, 53–79.

Chapter 3

1. Details of Nerilito Satur's death are based on a fact-finding report by the Multi-Sectoral Group, *The Heinous Killing of Rev. Fr. Nerilito Dazo Satur*, October 28, 1991; Carol Arguillas, "Fight for Environment Turns Bloody," *Philippine Daily Inquirer*, October 19, 1991; Carolyn O. Arguillas, "Murder amidst the Sunflowers," *Philippine Daily Inquirer*, November 5, 1991; John J. Carroll, "A Martyr for the Environment," *Philippine Daily Inquirer*, November 7, 1991; interviews in Valencia, Bukidnon, October, 1991.

2. Republic of the Philippines, *1994 Philippine Yearbook* (Manila: National Statistics Office, 1995), pp. 547, 549.

3. Catholic Bishops Conference of the Philippines, "What Is Happening to Our Beautiful Land," January 1988.

4. Marites Dañguilan Vitug, *Power from the Forest: The Politics of Logging* (Manila: Philippine Center for Investigative Journalism, 1993), p. 67.

5. Carroll, "A Martyr for the Environment."

6. Luzviminda Francisco, "The First Vietnam: The Philippine-American War, 1899–1902," in *The Philippines Reader: A History of Colonialism, Neocolonialism, Dictatorship, and Resistance*, ed. Daniel B. Schirmer and Stephen Rosskamm Shalom (Boston: South End Press, 1987).

7. Rodney J. Sullivan, *Exemplar of Americanism: The Philippine Career of Dean C. Worcester* (Ann Arbor: The University of Michigan Center for South and Southeast Asian Studies, 1991), pp. 81–82.

8. Jim Zwick, ed., *Mark Twain's Weapons of Satire: Anti-Imperialist Writings on the Philippine-American War* (Syracuse, N.Y.: Syracuse University Press, 1992), pp. xvii –xix, xxi.

9. Zwick, *Mark Twain's Weapons of Satire*, p. xviii.

10. Sullivan, *Exemplar of Americanism*, pp. 80–85.

11. Alfred J. Beveridge, "Our Philippine Policy," in Schirmer and Shalom, *The Philippines Reader*, pp. 23–26.

12. Dean C. Worcester, *The Philippines Past and Present*, new ed. (New York: Macmillan, 1930), p. 599.

13. Worcester, *The Philippines Past and Present*, p. 599.

14. George Perkins Marsh, *Man and Nature* (Cambridge, Mass.: Harvard University Press, 1965), pp. xvii–xx.

15. Donald Worster, *Nature's Economy: A History of Ecological Ideas* (Cambridge: Cambridge University Press, 1994), p. x; Marsh, *Man and Nature*, pp. 14–52.

16. Harold K. Steen, *The U.S. Forest Service: A History* (Seattle: University of Washington Press, 1976), pp. 9–20.

17. Steen, *The U.S. Forest Service*, pp. 47–102.

18. Food and Agriculture Organization (FAO) of the United Nations, *Review of Forest Management of Tropical Asia*, FAO Forestry Paper (Rome: FAO, 1989), p. 147.

19. Lewis E. Gleeck, Jr., *American Institutions in the Philippines (1848–1941)* (Manila: Historical Conservation Society, 1976), p. 222.

20. Richard P. Tucker, "Five Hundred Years of Tropical Forest Exploitation," in *Lessons of the Rainforest*, ed. Suzanne Head and Robert Heinzman (San Francisco: Sierra Club Books, 1990), pp. 48–50.

21. Tucker, "Five Hundred Years of Tropical Forest Exploitation," pp. 48–49.

22. Worcester, *The Philippines Past and Present*, p. 601.

23. Quoted in Gleeck, *American Institutions in the Philippines*, p. 279.

24. David M. Kummer, *Deforestation in the Postwar Philippines* (Quezon City and Chicago: Ateneo de Manila University Press and the University of Chicago Press, 1992), p. 21; Vitug, *Power from the Forest*, p. 25; Peter Dauvergne, *Shadows in the Forest: Japan and the Politics of Timber in Southeast Asia* (Cambridge, Mass.: MIT Press, 1997), pp. 133, 146.

25. Gleeck, *American Institutions in the Philippines*, pp. 223–237; Food and Agriculture Organization, "Demonstration and Training in Forest, Forest Range, and Watershed Management," United Nations Development Program, confidential document, 1970, p. 27.

26. Antonio G. M. La Viña, "The State of Community Based Forest Management in the Philippines and the Role of Local Governments" (unpublished manuscript, 1999), p. 2.

27. Kummer, *Deforestation in the Postwar Philippines*, particularly pp. 42–43, 59, 154.

28. Maria Concepcion Cruz, Carrie A. Meyer, Robert Repetto, and Richard Woodward, *Population Growth, Poverty, and Environmental Stress: Frontier Migration in the Philippines and Costa Rica* (Washington, D.C.: World Resources Institute, 1992), pp. 4–5.

29. Gleeck, *American Institutions in the Philippines*, pp. 223–227.

30. Danilo Balete, interview with the author, Chicago, Ill., August 31, 2000.

31. David Joel Steinberg, *The Philippines: A Singular and a Plural Place*, 4th ed. (Boulder, Colo.: Westview Press, 2000), p. 26.

32. Alfred W. McCoy, "'An Anarchy of Families': The Historiography of State and Family in the Philippines," in *An Anarchy of Families: State and Family in the Philippines*, ed. Alfred W. McCoy (Madison and Manila: Center for Southeast Asian Studies, University of Wisconsin-Madison, and Ateneo de Manila Press, 1994), p. 8.

33. Vitug, *Power from the Forest*, pp. 14–15; James K. Boyce, *The Philippines: The Political Economy of Growth and Impoverishment in the Marcos Era* (Honolulu: University of Hawaii Press and OECD Development Centre, 1993), pp. 234, 225.

34. Philip Hurst, *Rainforest Politics: Ecological Destruction in South-East Asia* (London: Zed Books, 1990), p. 188; Catherine Caufield, *In the Rainforest: Report from a Strange, Beautiful, Imperiled World* (Chicago: University of Chicago Press, 1991), pp. 162–166.

35. Caufield, *In the Rainforest*, p. 163.

36. Dauvergne, *Shadows in the Forest*, pp. 133–163, 170–173.

37. FAO, *Review of Forest Management*, p. 148.

38. Eduardo Tadem, Johnny Reyes, and Linda Susan Magno, *Showcases of Underdevelopment in Mindanao: Fishes, Forest, and Fruits* (Davao City, Philippines: Alternate Resource Center, 1984), pp. 96, 154–159.

39. Vitug, *Power from the Forest*, pp. 103–135.

Chapter 4

1. World Bank Environment Department, *Republic of the Philippines: Conservation of Priority Protected Areas*, project document, April 1994.

2. Mohan Munasinghe and Jeffrey McNeely, eds., *Protected Area Economics and Policy: Linking Conservation and Sustainable Development* (Washington, D.C.: World Bank and World Conservation Union [IUCN], 1994); David Western and R. Michael Wright, *Natural Connections: Perspectives in Community-Based Conservation* (Washington, D.C.: Island Press, 1994).

3. John Dixon and Paul B. Sherman, *Economics of Protected Areas: A New Look at Benefits and Costs* (Washington, D.C.: Island Press, 1990), pp. 12–15.

4. William Cronon, ed., *Uncommon Ground: Rethinking the Human Place in Nature* (New York: Norton, 1995).

5. Roderick Nash, *Wilderness and the American Mind* (New Haven, Conn.: Yale University Press, 1982), pp. 106–113.

6. Donald Worster, *Nature's Economy: A History of Ecological Ideas*, 2nd ed. (Cambridge: Cambridge University Press, 1994), pp. 417–419; quote from p. 418.

7. Edward O. Wilson, *Biophilia* (Cambridge, Mass.: Harvard University Press, 1984), pp. 1, 121, 138–139.

8. Worster, *Nature's Economy*, p. 417.

9. World Conservation Union, available at www.iucn.org/themes/wcpa/wpc2003/images/unlist/26.pdf.

10. Norman Myers, "Threatened Biotas: 'Hot Spots' in Tropical Forests," *The Environmentalist*, 8, no. 3 (1988), pp. 187–208.

11. Eric Dinerstein and Eric D. Wikramanayake, "Beyond 'Hotspots': How to Prioritize Investments to Conserve Biodiversity in the Indo-Pacific Region," *Conservation Biology* 7, no. 1 (March 1993), pp. 53–65.

12. Larry Heaney, telephone conversations and interviews with the author in Chicago, Field Museum of Natural History, 1999, 2000, 2004.

13. David Western, quoted in Primack, *Primer of Conservation Biology*, p. 196.

14. Western and Wright, *Natural Connections*; Richard Primack, *Primer of Conservation Biology* (Sunderland, Mass.: Sinauer Associates, 1995), pp. 164–194.

15. Danilo Balete, interviews with the author in Chicago, Field Museum of Natural History, August 2000.

16. Richard Anson, telephone conversations with the author, 2003 and 2004.

Chapter 5

1. E. D. Gomez, "Reef Management in Developing Countries: A Case Study in the Philippines," *Coral Reefs* 16, suppl. (1997), pp. S3–S8; Alan T. White and Annabelle Cruz-Trinidad, *The Values of Philippine Coastal Resources: Why Protection and Management Are Critical* (Cebu City, Philippines: Coastal Resource Management Project, 1998), pp. 1–2; Alan White, personal communication with the author, July 2004; Asian Development Bank, "Fisheries Sector Development Project, Philippines" (draft final report, Manila, 1996), p. 35.

2. In the Philippines, the terms *marine protected areas, reserves*, and *sanctuaries* are used synonymously.

3. Asian Development Bank, "Fisheries Sector Development Project," pp. 22–23.

4. Silliman University Marine Laboratory, Apo Island information sheet, 1995.

5. Santos B. Rasalan, "Marine Fisheries of the Central Visayas," *The Philippine Journal of Fisheries* 5, no. 1 (1957), p. 86.

6. Republic of the Philippines, *The Philippine Economic Atlas* (Manila: Philippine Office of the President, [1964?]).

7. Charles Victor Barber and Vaughan R. Pratt, *Sullied Seas: Strategies for Combating Cyanide Fishing in Southeast Asia and Beyond* (Washington, D.C., and Manila: World Resources Institute and the International MarineLife Alliance, 1997), pp. vii, 25–39.

8. Rasalan, "Marine Fisheries," pp. 85–86.

9. Garry R. Russ and Angel C. Alcala, "Marine Reserves: Rates and Patterns of Recovery and Decline of Large Predatory Fish," *Ecological Applications* 6, no. 3 (1996), pp. 947–961.

10. T. H. Huxley, "The Herring," *Nature* (April 28, 1881), 607–613.

11. Harden F. Taylor, *Survey of Marine Fisheries of North Carolina* (Chapel Hill: University of North Carolina Press, 1951), pp. 312–315.

12. Food and Agriculture Organization (FAO) of the United Nations, FAO Fisheries Department, *The State of World Fisheries and Aquaculture* (Rome: United Nations, 1996), pp. 3–30.

13. Daniel Pauly, Villy Christensen, Sylvie Guénette, Tony J. Pitcher, U. Rashid Sumaila, Carl J. Walters, R. Watson, and Dirk Zeller, "Towards Sustainability in World Fisheries," *Nature* 418 (August 8, 2002), pp. 689–695.

14. Ricardo P. Babaran, Jose A. Ingles, and Teodoro U. Abalos, "Defining the University's Agenda for the Fisheries Sector," in *Archipelagic Studies: Charting New Waters*, ed. Jay L. Batongbacal (Quezon City: University of the Philippines, 1998), p. 66.

15. James A. Bohnsack, "Marine Reserves: They Enhance Fisheries, Reduce Conflicts, and Protect Resources," *Oceanus* (Fall 1993), pp. 64–65.

16. Peter F. Sale, Graham E. Forrester, and Phillip S. Levin, "Reef Fish Management," *National Geographic Research and Exploration* 10, no. 2 (1994), pp. 225–226, 228.

17. Jennifer E. Caselle and Robert R. Warner, "Variability in Recruitment of Coral Reef Fishes: The Importance of Habitat at Two Spatial Scales," *Ecology* 77, no. 8 (December 1996), pp. 2488–2489.

18. Russ and Alcala, "Marine Reserves," p. 948.

19. Garry R. Russ and Angel C. Alcala, "Do Marine Reserves Export Adult Fish Biomass? Evidence from Apo Island, Central Philippines," *Marine Ecology Progress Series* 132 (1996), pp. 1–9.

20. Garry R. Russ and Angel C. Alcala, "Effects of Intense Fishing Pressure on an Assemblage of Coral Reef Fishes," *Marine Ecology Progress Series* 56 (1989), pp. 13–27.

21. Angel C. Alcala, "Effects of Marine Reserves on Coral Fish Abundances and Yields of Philippine Coral Reefs," *Ambio* 17, no. 3 (1988), pp. 194–199; Robert S. Pomeroy and Melvin B. Carlos, "Community-Based Coastal Management in the Philippines: A Review and Evaluation of Programs and Projects, 1984–1994," *Marine Policy* 21, no. 5 (1997), pp. 445–464.

22. Alan T. White, personal communications with the author, Cebu City, July 1999, July 2004.

Chapter 6

1. Details of the Bolinao case were drawn from interviews in Manila and Bolinao; interviews by the author with Antonio La Viña in 1996, 2000, and 2004; and documents from the Pangasinan Cement Corporation.

2. Sally Ness, *Where Asia Smiles: An Ethnography of Philippine Tourism* (Philadelphia: University of Pennsylvania Press, 2003), pp. 108–114.

3. Republic of the Philippines, *Medium Term Development Plan: 1993–1998* (Manila: National Economic and Development Authority, March 1995), pp. 1–9.

4. *BusinessWorld*, July 27, 1996, pp. 5–9.

5. Robert McCaffrey, "Storm Clouds Ahead—Or Not?" *Asian Cement and Construction Materials Magazine*, February 1999; DENR, "List of Approved Mineral Production Sharing Agreements," updated as of June 30, 2000.

6. "A Proposed Cement Plant Complex in Bolinao, Pangasinan: Potential Conflicts in the Coastal Zone," position paper endorsed by the University of the Philippines Diliman Executive Committee and the University Council on January 21, 1995; Elmer Ferrer and Emmanuel M. Luna, "Nurturing the Seeds for Action: The Bolinao Cement Plant Controversy as a Case for the Academe's Involvement in Social Issues," *Philippine Democracy Agenda* (n.p., n.d.), pp. 205–220.

7. Victor O. Ramos, Letter from the Secretary of the Department of Environment and Natural Resources to Raymundo T. Gonzales, Director of Pangasinan Cement Corporation, October 30, 1995.

8. Catherine Coumans, "Canadian Transnational Corporation Dumps Waste, Responsibility in Marinduque" (Manila: Philippine Center for Investigative Journalism, March 24–26, 1999). Available at www.pcij.org/stories/1999/marcopper.html (accessed June 15, 2005).

9. Emma Helen Blair and James Alexander Robertson, *The Philippine Islands, 1493–1898*, Volume X, *1597–1599* (Cleveland, Ohio: Arthur Clark, 1904).

10. Paul Hutchcroft, *Booty Capitalism: The Politics of Banking in the Philippines* (Ithaca, N.Y.: Cornell University Press, 1998), pp. 13–30.

11. Candido Cabrido, interviews with the author, 1996 and 1997.

12. Lingayen Gulf Coastal Area Management Commission, "A Brief on the Commission and the Gulf," n.d. In *Lingayen Gulf: God's Gift, Our Heritage, Our Responsibility.*

13. "A Proposed Cement Plant Complex in Bolinao, Pangasinan: Potential Conflicts in the Coastal Zone," position paper endorsed by the University of the Philippines Diliman Executive Committee and the University Council on January 21, 1995; Fidel V. Ramos, "Proclaiming Lingayen Gulf as an Environmentally Critical Area," Presidential Proclamation No. 156, March 25, 1993.

14. Victor O. Ramos, letter to Andrew E. J. Wang, General Manager, Pangasinan Cement Corporation, Quezon City, DENR, August 6, 1996.

Chapter 7

1. Ricardo M. Sandalo, "Sustainable Development and the Environmental Plan for Palawan," in *Palawan at the Crossroads: Development and the Environment on a Philippine Frontier*, ed. James F. Eder and Janet O. Fernandez (Quezon City: Ateneo de Manila University Press, 1996), pp. 127–128.

2. Sally Ness, *Where Asia Smiles: An Ethnography of Philippine Tourism* (Philadelphia: University of Pennsylvania Press, 2003), pp. 108–125.

3. Figures from Haribon Palawan, 1991.

4. Japan International Cooperation Agency (JICA) and the Department of Tourism, Philippines, "The Study on Environmentally Sustainable Tourism Development Plan for Northern Palawan in the Republic of the Philippines," progress report, March 1996, p. S-5.

5. Quoted in Palawan Council for Sustainable Development, Office of the President, *Strategic Environmental Plan for Palawan: Towards Sustainable Development* (Manila: Palawan Integrated Area Development Project Office, [1989?]), p. 1.

6. Palawan Council for Sustainable Development, *Strategic Environmental Plan*, pp. 4–5.

7. Yasmin D. Arquiza, "Palawan's Environmental Movement," in Eder and Fernandez, *Palawan at the Crossroads*, pp. 136-145.

8. Celso R. Roque, *Earth, Water, Air, Fire: Essays on Environmentalism* (Manila: Kalikasan Press, 1990), p. 21.

9. *Philippine Daily Inquirer*, March 8, 1991.

10. JICA and the Department of Tourism, "Study on Environmentally Sustainable Tourism," pp. S-7, 2.48, 2.54.

11. Gregor Hodgson and John A. Dixon, "Logging Versus Fisheries in the Philippines," *The Ecologist* 19, no. 4 (1989), pp. 138-143.

12. Yehuda Amichai, *The Selected Poetry of Yehuda Amichai*, ed. and trans. Chana Bloch and Stephen Mitchell (Berkeley: University of California Press, 1996), pp. 137-138.

13. Dean MacCannell, *The Tourist: A New Theory of the Leisure Class* (Berkeley: University of California Press, 1999), pp. 11-13.

14. David A. Fennell, *Ecotourism: An Introduction* (London: Routledge, 1999), p. 100.

15. Motoe Terami-Wada, "Karayuki-san of Manila: 1890-1920," *Philippine Studies* 34 (1986), pp. 287-316; quoted in Saundra Pollock Sturdevant and Brenda Stoltzfus, *Let the Good Times Roll: Prostitution and the U.S. Military in Asia* (New York: New Press, 1993), p. 303.

16. Allan M. Brandt, *No Magic Bullet: A Social History of Venereal Disease in the United States since 1880* (New York: Oxford University Press, 1987), pp. 52-70.

17. Linda K. Richter, *The Politics of Tourism in Asia* (Honolulu: University of Hawaii Press, 1989), pp. 51-81.

18. Thanh-Dam Truong, *Sex, Money, and Morality: Prostitution and Tourism in Southeast Asia* (London: Zed Books, 1990), p. 128; Richter, *The Politics of Tourism in Asia*, p. 78.

19. JICA and the Department of Tourism, "Study on Environmentally Sustainable Tourism," pp. 2.114, 3.60.

20. Redempto Anda and Grizelda (Gerthie) Mayo-Anda, interviews with author, 1996, 1997.

21. JICA and the Department of Tourism, "Study on Environmentally Sustainable Tourism," pp. 2.49-2.51.

22. *The Palawan Times*, No. 329, April 2-8, 2001.

23. Gregor Hodgson, phone interview with the author, 2001.

24. Wilbur Dee, Biodiversity in Development Project, "Case Study: El Nido Managed Resource Protected Area," Department for International Development, European Commission, and the World Conservation Union (n.d.), section 7.1.2. Available at www.wcmc.org.uk/biodev/case%20study/philippines.pdf.

25. Criselda Yabes, "The North and South of Palawan: The Loss of a Paradise," *Bandillo ng Palawan—The Towncrier of Palawan*, June 1999.

Chapter 8

1. Some details of the goals and accomplishments of the Comprehensive Agrarian Reform Program were taken from a 2003 internal study by the Department of Agrarian Reform.

2. Saturnino M. Borras, *The Bibingka Strategy in Land Reform Implementation: Autonomous Peasant Movements and State Reformists in the Philippines* (Quezon City: Institute for Popular Democracy, 1998), pp. 1–5.

3. Jennifer C. Franco, "Between Uncritical Collaboration and Outright Opposition: An Evaluation Report on the Partnership for Agrarian Reform and Rural Development Services and the Struggle for Agrarian Reform and Rural Development in the 1990s" (Quezon City: Institute for Popular Democracy, 1998).

4. Stephen Leonides, provincial agrarian reform officer, Negros Occidental, interviews with author, August 1997 and August 1999, Bacolod, Philippines. Other details are based on interviews conducted in Bacolod, Baviera, and Manila.

5. Yujiro Hayami, Ma. Agnes R. Quisumbing, and Lourdes S. Adriano, *Toward an Alternative Land Reform Paradigm: A Philippine Perspective* (Quezon City: Ateneo de Manila Press, 1990), p. 26; Michael S. Billig, "Syrup in the Wheels of Progress: The Inefficient Organization of the Philippine Sugar Industry," *Journal of Southeast Asian Studies* 24 (March 1993), pp. 122–147.

6. James Putzel, *A Captive Land: The Politics of Agrarian Reform in the Philippines* (Quezon City: Ateneo de Manila University Press, 1992), pp. 20–29.

7. Putzel, *A Captive Land*, p. 28.

8. Hayami, Quisumbing, and Adriano, *Toward an Alternative Land Reform Paradigm*, pp. 28–33.

9. Republic of the Philippines, *Handbook for CARP Implementors*, 2nd ed. (Quezon City: Department of Agrarian Reform, Bureau of Agrarian Reform Information and Education, 1995), p. 4.

10. O. D. Corpuz, *An Economic History of the Philippines* (Quezon City: University of the Philippines, 1997), pp. 2, 34, 133, 150–151.

11. Corpuz, *An Economic History*, p. 111.

12. Michael S. Billig, "The Rationality of Growing Sugar in Negros," *Philippine Studies* 40 (1992), p. 156.

13. Hayami, Quisumbing, and Adriano, *Toward an Alternative Land Reform Paradigm*, p. 40.

14. Alan Berlow, *Dead Season: A Story of Murder and Revenge on the Philippine Island of Negros* (New York: Pantheon Books, 1996), p. 76.

15. Temario C. Rivera, *Landlords and Capitalists: Class, Family, and State in Philippine Manufacturing* (Quezon City: University of the Philippines Press, 1994), pp. 125–130.

16. Berlow, *Dead Season*, p. 92.

17. Corpuz, *An Economic History*, pp. 14–18.

18. Corpuz, *An Economic History*, pp. 24–29.

19. Corpuz, *An Economic History*, pp. 61–65; Hayami, Quisumbing, and Adriano, *Toward an Alternative Land Reform Paradigm*, p. 45.

20. David Joel Steinberg. *The Philippines: A Singular and a Plural Place*, 4th ed. (Boulder, Colo.: Westview Press, 2000), pp. 25–26.

21. Putzel, *A Captive Land*, p. 58.

22. Putzel, *A Captive Land*, pp. 58, 96–105, 153–156; Hayami, Quisumbing, and Adriano, *Toward an Alternative Land Reform Paradigm*, pp. 25, 62–70.

23. Statement by Council of Agricultural Producers of the Philippines; quoted in Putzel, *A Captive Land*, p. 1.

24. Corazon Aquino, selection from campaign speech delivered in Davao City, Mindanao, on January 16, 1986; reprinted in *The Philippines Reader: A History of Colonialism, Neocolonialism, Dictatorship, and Resistance*, ed. Daniel B. Schirmer and Stephen Rosskamm Shalom (Boston: South End Press, 1987), p. 339.

25. Putzel, *A Captive Land*, pp. 181–185.

26. Putzel, *A Captive Land*, pp. 198, 199, 213.

27. Borras, *The Bibingka Strategy*, p. 54.

28. Putzel, *A Captive Land*, p. 275.

29. Billig, "The Rationality of Growing Sugar," p. 153.

30. Philippine Campaign for Agrarian Reform and Rural Development (Phil-CARRD), *Proceedings of the Agrarian Reform and Rural Development National Consultation* (Quezon City: Partnership for Agrarian Reform and Rural Development Services, Inc., June 5–6, 1995), p. 24.

31. Gerardo Bulatao, PhilCARRD, 1995, p. 114.

32. Republic of the Philippines, *1994 Philippine Yearbook* (Manila: National Statistics Office, 1995), pp. 132–136, tables 5.1, 5.2, and 5.6.

Chapter 9

1. Marcia D. Lowe, "Shaping Cities: The Environmental and Human Dimensions," Worldwatch Paper 105 (Washington, D.C.: Worldwatch Institute, 1991), p. 6; G. Thomas Kingsley, Bruce W. Ferguson, and Blair T. Bower with Stephen R. Dice, "Managing Urban Environmental Quality in Asia," World Bank Technical Paper No. 200 (Washington, D.C.: World Bank, 1994), pp. 1–2; Ismail Serageldin, Michael A. Cohen, and K. D. Sivaramakrishnan, "The Human Face of the Urban Environment," *Proceedings of the Second Annual World Bank Conference on Environmentally Sustainable Development* (Washington, D.C.: World Bank, 1994), p. 21.

2. World Bank, *Urban Air Quality Management Strategy in Asia: Metro Manila Report*, World Bank Technical Paper No. 380 (Washington, D.C.: World Bank, 1997), p. 148.

3. George Esguerra, Assistant Secretary for Planning, Department of Transportation and Communications, interview with the author, August 1999.

4. Republic of the Philippines, "Metro Manila Urban Transportation Integration Study" (draft final report, Manila, December 1998), p. S-9; World Bank, *Philippines Environment Monitor* (Washington, D.C., and Manila: World Bank, 2002 and 2004); Ronald Subida and Dan MacNamara, "Metropolitan Manila: A Focus on the Transport Sector," Integrated Environmental Strategies Philippines Project Report, Manila Observatory and USAID, June 2005, available at www.epa.gov/ies/documents/philippines/ies_update.pdf

5. Philippine Department of Health, Health Indicators, available at www.doh.gov.ph/healthsit/health_indicators.htm (accessed July 2004).

6. Elisea (Bebet) Gozun, interviews with the author, Quezon City, August 1999.

7. Ruby Paredes, "The Pardo de Taveras of Manila," in *An Anarchy of Families: State and Family in the Philippines*, ed. Alfred W. McCoy (Madison and Manila: Center for

Southeast Asian Studies, University of Wisconsin-Madison and Ateneo de Manila University Press, 1994), pp. 365–366 and p. 420, n. 51.

8. United States Bureau of Insular Affairs, War Department, *Reports of the Philippine Commission: 1900–1903* (Washington, D.C.: Government Printing Office, 1904), 1901 report, pp. 139, 152–153.

9. Daniel Burnham, "The Development of Manila," *The Western Architect*, January 1906, p. 7; quoted in David Brody, "Fantasy Realized: The Philippines, Orientalism, and Imperialism in Turn-of-the-Century American Culture" (Ph.D. diss., Boston University, May 1996), p. 12; Jane Jacobs, *The Death and Life of Great American Cities* (New York: Vintage Books, 1961), pp. 24–25; Thomas S. Hines, "The Imperial Mall: The City Beautiful Movement and the Washington Plan, 1901–1902," *Studies in the History of Art* 30 (1991), p. 79.

10. Rosario D. Jimenez and Sister Aida Velasquez, "Metropolitan Manila: A Framework for Its Sustained Development," *Environment and Urbanization* 1, no. 1 (April 1989), p. 52.

11. Jacobs, *Death and Life of Great American Cities*, pp. 16, 428–448.

12. Alan Altshuler and David Luberoff, *Mega-Projects: The Changing Politics of Urban Public Investment* (Washington, D.C.: The Brookings Institution Press, 2003), pp. 22–26.

13. Maria (Fides) Bagasao, interim executive director, CO-TRAIN, interview with the author, August 1997.

14. Corazon (Dinky) Juliano Soliman, quoted in the Bulletin Archives, Kennedy School of Government, Harvard University, Summer/Fall 1999. Soliman was later appointed secretary of the Department of Social Welfare and Development.

15. CO-TRAIN, *Lessons and Tradition of 20 Years Experience* (Quezon City: CO-TRAIN, n.d.).

16. Interviews by the author with officials at the River Rehabilitation Secretariat, DENR, Quezon City, August 1999.

17. Joel Rocamora, "Classes, Bosses, Goons, and Clans: Re-imagining Philippine Political Culture," in *Boss: Five Case Studies of Local Politics in the Philippines*, ed. Jose F. Lacaba (Manila: Philippine Center for Investigative Journalism and Institute for Popular Democracy, 1995), p. xiii.

18. Arsenio M. Balisacan, *Poverty, Urbanization and Development Policy: A Philippine Perspective* (Quezon City: University of the Philippines Press, 1994), p. 62, 136–140.

19. Philippine National Statistics Office, "Survey on Overseas Filipinos," available at www.census.gov.ph (accessed August 2004); Barbara Goldoftas, "Despite Gulf Crisis, Migrant Filipinas Still Pin Hopes on Petrodollar," *Isis* 1 (1991), pp. 12–14; Seth Mydans, "Looking Out for the Many, in Saving the One," *New York Times*, August 1, 2004, p. 3.

20. Details drawn from interviews with Antonio Oposa and Maynilad Water, Manila Water, MWSS, and Asian Development Bank officials who prefer to remain anonymous. Details about the MWSS privatization are also based on Arthur C. McIntosh, *Asian Water Supplies: Reaching the Urban Poor* (Manila: Asian Development Bank, 2003); available at www.adb.org/Documents/Books/Asian_Water_Supplies.

21. *Oposa et al v. Fulgencio S. Factoran*, 1993.

22. David Llorito, "Maynilad: A Model in Water Privatization Springs Leaks," *Manila Times* special report, March 26, 2003; available at www.manilatimes.net.

23. Luz Rimban, "Big Companies Steal Water," Philippine Center for Investigative Journalism, March 6–7, 2000; available at www.pcij.org/stories/2000/water.html.

24. William Finnegan, "Leasing the Rain," *New Yorker*, August 4, 2002, pp. 43–53.

25. José A. Gómez-Ibáñez, "Private Infrastructure and the Search for Commitment," Harvard University Kennedy School of Government Taubman Center for State and Local Government Annual Report, 2003; available at www.ksg.harvard.edu/taubmancenter.

26. Public Citizen, "Suez: A Corporate Profile" (Washington, D.C.: Public Citizen, April 2005); available at www.citizen.org/documents/profilesuez.pdf .

27. Shane Rosenthal, "The Manila Water Concessions and Their Impact on the Poor," Yale School of Forestry and Environmental Studies, 2001. Available at www.yale.edu/hixon/research/pdf/Srosenthal_Malina.pdf.

28. Untitled Manila Water Company documents, data on sewerage coverage targets, 1999.

Epilogue

1. J. R. McNeill, *Something New under the Sun: An Environmental History of the Twentieth-Century World* (New York: Norton, 2000), pp. 287–289.

2 World Bank, *Philippines Environment Monitor* (Washington, D.C., and Manila: World Bank, 2004).

Bibliography

Alcala, Angel C. "Effects of Marine Reserves on Coral Fish Abundances and Yields of Philippine Coral Reefs." *Ambio* 17, no. 3 (1988): 194–199.

Altshuler, Alan, and David Luberoff. *Mega-Projects: The Changing Politics of Urban Public Investment*. Washington, D.C.: Brookings Institution Press, 2003.

Amichai, Yehuda. *The Selected Poetry of Yehuda Amichai*. Edited and translated by Chana Bloch and Stephen Mitchell. Berkeley: University of California Press, 1996.

Asian Development Bank. "Fisheries Sector Development Project, Philippines." Draft final report, Manila, 1996.

Asian Development Bank. "Manila Air Quality Improvement Sector Development Program." Report and recommendation, Manila, November 1998.

Balisacan, Arsenio. *The Philippine Economy: Development, Policies, and Challenges*. New York: Oxford University Press, 2002.

Balisacan, Arsenio. *Poverty, Urbanization and Development Policy: A Philippine Perspective*. Quezon City: University of the Philippines Press, 1994.

Barber, Charles Victor, and Vaughan R. Pratt. *Sullied Seas: Strategies for Combating Cyanide Fishing in Southeast Asia and Beyond*. Washington, D.C., and Manila: World Resources Institute and the International MarineLife Alliance, 1997.

Bates, David V. "The Effects of Air Pollution on Children." *Environmental Health Perspectives* 103, suppl. 6 (September 1995): 49–53.

Batongbacal, Jay L., ed. *Archipelagic Studies: Charting New Waters*. Quezon City: University of the Philippines, 1998.

Bello, Walden, Shea Cunningham, and Li Kheng Poh. *A Siamese Tragedy: Development and Disintegration in Modern Thailand*. London: Zed Books, 1998.

Berlow, Alan. *Dead Season: A Story of Murder and Revenge on the Philippine Island of Negros*. New York: Pantheon Books, 1996.

Berman, Stephen. "Epidemiology of Acute Respiratory Infections in Children of Developing Countries." *Review of Infectious Diseases* 13, suppl. 6 (1991): S454–S462.

Billig, Michael S. "The Rationality of Growing Sugar in Negros." *Philippine Studies* 40 (1992): 153–182.

Billig, Michael S. "Syrup in the Wheels of Progress: The Inefficient Organization of the Philippine Sugar Industry." *Journal of Southeast Asian Studies* 24 (March 1993): 122–147.

Blair, Emma Helen, and James Alexander Robertson. *The Philippine Islands, 1493–1898.* Cleveland, Ohio: Arthur Clark, 1904.

Bohnsack, James A. "Marine Reserves: They Enhance Fisheries, Reduce Conflicts, and Protect Resources." *Oceanus* (Fall 1993): 64–65.

Bonner, Raymond. *Waltzing with a Dictator: The Marcoses and the Making of American Policy.* New York: Vintage Books, 1988.

Borras, Saturnino M. *The Bibingka Strategy in Land Reform Implementation: Autonomous Peasant Movements and State Reformists in the Philippines.* Quezon City: Institute for Popular Democracy, 1998.

Boyce, James K. *The Philippines: The Political Economy of Growth and Impoverishment in the Marcos Era.* Honolulu: University of Hawaii Press and OECD Development Centre, 1993.

Brandt, Allan M. *No Magic Bullet: A Social History of Venereal Disease in the United States since 1880.* New York: Oxford University Press, 1987.

Brechin, Stephen. *Planting Trees in the Developing World: A Sociology of International Organizations.* Baltimore, Md.: Johns Hopkins University Press, 1997.

Broad, Robin, and John Cavanagh. *The Philippine Challenge: Sustainable and Equitable Development in the 1990s.* Quezon City: Philippine Center for Policy Studies, 1991.

Broad, Robin, with John Cavanagh. *Plundering Paradise: The Struggle for the Environment in the Philippines.* Berkeley: University of California Press, 1993.

Brody, David. "Fantasy Realized: The Philippines, Orientalism, and Imperialism in Turn-of-the-Century American Culture." Ph.D. dissertation, Boston University, 1996.

Burnham, D. H. "Report on Proposed Improvements at Manila." From the Report of the Philippine Commission, Part 1, pp. 627–635. Washington, D.C.: Bureau of Insular Affairs, War Department, 1906.

Caselle, Jennifer E., and Robert R. Warner. "Variability in Recruitment of Coral Reef Fishes: The Importance of Habitat at Two Spatial Scales." *Ecology* 77, no. 8 (December 1996): 2488–2504.

Catholic Bishops Conference of the Philippines. "What Is Happening to Our Beautiful Land." January 1988.

Caufield, Catherine. *In the Rainforest: Report from a Strange, Beautiful, Imperiled World.* Chicago: University of Chicago Press, 1991.

Cesar, Herman, S. J. *Collected Essays on the Economics of Coral Reefs.* Kalmar, Sweden: Kalmar University, 2000.

Christie, Patrick, ed. Theme Issue: Tropical Coastal Management. *Coastal Management* 28, no. 1 (January–March 2000).

Coastal Resource Management Project. *Philippine Coastal Resource Management Guidebook Series.* 8 vols. www.oneocean.org.

Constantino, Renato, and Letizia Constantino. *The Philippines: The Continuing Past.* Quezon City: The Foundation for Nationalist Studies, 1978.

Coronel, Sheila S., ed. *Patrimony: Six Case Studies on Local Politics and the Environment in the Philippines.* Manila: Philippine Center for Investigative Journalism, 1996.

Corpuz, O. D. *An Economic History of the Philippines.* Quezon City: University of the Philippines Press, 1997.

Cronon, William, ed. *Uncommon Ground: Rethinking the Human Place in Nature.* New York: Norton, 1995.

Cruz, Maria Concepcion, Carrie A. Meyer, Robert Repetto, and Richard Woodward. *Population Growth, Poverty, and Environmental Stress: Frontier Migration in the Philippines and Costa Rica.* Washington, D.C.: World Resources Institute, 1992.

Daily, Gretchen C., ed. *Nature's Services: Societal Dependence on Natural Ecosystems.* Washington, D.C.: Island Press, 1997.

Darwin, Charles. *The Voyage of the Beagle.* New York: Bantam Books, 1958.

Dauvergne, Peter. *Loggers and Degradation in the Asia-Pacific: Corporations and Environmental Management.* Cambridge: Cambridge University Press, 2001.

Dauvergne, Peter. *Shadows in the Forest: Japan and the Politics of Timber in Southeast Asia.* Cambridge, Mass.: MIT Press, 1997.

Devas, Nick, and Carole Rakodi. *Managing Fast Growing Cities: New Approaches to Urban Planning and Management in the Developing World.* Essex, England: Longman Scientific and Technical, 1993.

Dickey, Jefferson H. *No Room to Breathe: Health Effects of Criteria Air Pollutants from Power Plants.* Boston: Greater Boston Physicians for Social Responsibility, 2004. Available at www.igc.org.

Dinerstein, Eric, and Eric D. Wikramanayake. "Beyond 'Hotspots': How to Prioritize Investments to Conserve Biodiversity in the Indo-Pacific Region." *Conservation Biology* 7, no. 1 (March 1993): 53–65.

Dixon, John, and Paul B. Sherman. *Economics of Protected Areas: A New Look at Benefits and Costs.* Washington, D.C.: Island Press, 1990.

Doeppers, Daniel F., and Peter Xenos. *Population and History: The Demographic Origins of the Modern Philippines.* Quezon City: Ateneo de Manila University Press, published in cooperation with University of Wisconsin-Madison Center for Southeast Asian Studies, 1998.

Drakakis-Smith, David. *The Nature of the Third World Cities.* Copenhagen: Center for Udviklingsforskning/Centre for Development Research, 1993.

Eder, James F., and Janet O. Fernandez, eds. *Palawan at the Crossroads: Development and the Environment on a Philippine Frontier.* Quezon City: Ateneo de Manila Press, 1996.

Esquillo, Ruth M. "Community Action on Forest Protection: The Case of San Fernando, Bukidnon." Master's thesis, Ateneo de Manila University, Quezon City, 1992.

Fallows, James. *Looking at the Sun: The Rise of the New East Asian Economic and Political System.* New York: Vintage Books, 1995.

Fennell, David A. *Ecotourism: An Introduction.* London: Routledge, 1999.

Ferrer, Elmer, and Emmanuel M. Luna. "Nurturing the Seeds for Action: The Bolinao Cement Plant Controversy as a Case for the Academe's Involvement in Social Issues." *Philippine Democracy Agenda.* n.d.

Food and Agriculture Organization (FAO) of the United Nations. "Demonstration and Training in Forest, Forest Range, and Watershed Management." United Nations Development Program, confidential document, 1970.

Food and Agriculture Organization (FAO) of the United Nations. *Review of Forest Management of Tropical Asia*. FAO Forestry Paper. Rome: United Nations, 1989.

Food and Agriculture Organization (FAO) of the United Nations. *The State of World Fisheries and Aquaculture*. FAO Fisheries Department. Rome: United Nations, 1996.

Franco, Jennifer C. "Between Uncritical Collaboration and Outright Opposition: An Evaluation Report on the Partnership for Agrarian Reform and Rural Development Services and the Struggle for Agrarian Reform and Rural Development in the 1990s." Quezon City: Institute for Popular Democracy, 1998.

Fuchs, Lawrence H. *"Those Peculiar Americans": The Peace Corps and the American National Character*. New York: Meredith Press, 1967.

Gleeck, Jr., Lewis E. *American Institutions in the Philippines (1848–1941)*. Manila: Historical Conservation Society, 1976.

Gleick, Peter. *The World's Water: The Biennial Report on Freshwater Resources: 2004–2005*. Washington, D.C.: Island Press, 2004.

Goldoftas, Barbara. "Despite Gulf Crisis, Migrant Filipinas Still Pin Hopes on Petrodollar." *Isis* 1 (1991): 12–14.

Gomez, E. D. "Reef Management in Developing Countries: A Case Study in the Philippines." *Coral Reefs* 16, suppl. (1997): 3–8.

Hayami, Yujiro, Ma. Agnes R. Quisumbing, and Lourdes S. Adriano. *Toward an Alternative Land Reform Paradigm: A Philippine Perspective*. Quezon City: Ateneo de Manila Press, 1990.

Head, Suzanne, and Robert Heinzman, eds. *Lessons of the Rainforest*. San Francisco: Sierra Club Books, 1990.

Heaney, Lawrence R., and Jacinto C. Regalado, Jr. *Vanishing Treasures of the Philippine Rain Forest*. Chicago: Field Museum, 1998.

Hillel, Daniel J. *Out of the Earth: Civilization and the Life of the Soil*. New York: Free Press, 1991.

Hines, Thomas S. "The Imperial Mall: The City Beautiful Movement and the Washington Plan, 1901–1902." *Studies in the History of Art* 30 (1991): 78–99.

Hodgson, Gregor, and John A. Dixon. "Logging Versus Fisheries in the Philippines." *The Ecologist* 19, no. 4 (1989): 138–143.

Howard, Michael C., ed. *Asia's Environmental Crisis*. Boulder, Colo.: Westview Press, 1993.

Humboldt, Alexander von. *Aspects of Nature in Different Lands and Different Climates, with Scientific Elucidations*. Translated by Mrs. Sabine. Vol. 1. London: Longman, Brown, Green, and Longmans, 1849.

Hurst, Philip. *Rainforest Politics: Ecological Destruction in South-East Asia*. London: Zed Books, 1990.

Hutchcroft, Paul. *Booty Capitalism: The Politics of Banking in the Philippines*. Ithaca, N.Y.: Cornell University Press, 1998.

Huxley, T. H. "The Herring." *Nature* 28 (April 1881): 607–613.

Jabla, Jun. *Defending the Forest: A Case Study of San Fernando, Bukidnon, Philippines.* Davao, Mindanao, Philippines: Kinaiyahan Foundation, 1990.

Jacobs, Jane. *The Death and Life of Great American Cities.* New York: Vintage Books, 1961.

Japan International Cooperation Agency (JICA) and the Department of Tourism, Philippines. "The Study on Environmentally Sustainable Tourism Development Plan for Northern Palawan in the Republic of the Philippines." Progress report, March 1996.

Jimenez, Rosario D., and Sister Aida Velasquez. "Metropolitan Manila: A Framework for Its Sustained Development." *Environment and Urbanization* 1, no. 1 (April 1989).

Kaplan, Robert D. *The Ends of the Earth.* New York: Vintage Books, 1997.

Karnow, Stanley. *In Our Image: America's Empire in the Philippines.* New York: Ballantine Books, 1989.

Kessler, Richard J. *Rebellion and Repression in the Philippines.* New Haven, Conn.: Yale University Press, 1989.

Kingsley, G. Thomas, Bruce W. Ferguson, and Blair T. Bower with Stephen R. Dice. "Managing Urban Environmental Quality in Asia." World Bank Technical Paper No. 200. Washington, D.C.: World Bank, 1994.

Kummer, David M. *Deforestation in the Postwar Philippines.* Quezon City and Chicago: Ateneo de Manila University Press and the University of Chicago Press, 1992.

Lacaba, Jose F., ed. *Boss: Five Case Studies of Local Politics in the Philippines.* Manila: Philippine Center for Investigative Journalism and Institute for Popular Democracy, 1995.

Lowe, Marcia D. "Shaping Cities: The Environmental and Human Dimensions." Worldwatch Paper 105. Washington, D.C.: Worldwatch Institute, 1991.

MacCannell, Dean. *The Tourist: A New Theory of the Leisure Class.* Berkeley: University of California Press, 1999.

Marsh, George Perkins. *Man and Nature.* Cambridge, Mass.: Harvard University Press, 1965.

McCoy, Alfred W., ed. *An Anarchy of Families: State and Family in the Philippines.* Madison and Manila: Center for Southeast Asian Studies, University of Wisconsin-Madison, and Ateneo de Manila Press, 1994.

McGoodwin, James R. *Crisis in the World's Fisheries: People, Problems, and Policies.* Palo Alto, Calif.: Stanford University Press, 1990.

McIntosh, Arthur C. *Asian Water Supplies: Reaching the Urban Poor.* Manila: Asian Development Bank, 2003. Available at www.adb.org/Documents/Books/Asian_Water_Supplies.

McNeill, J. R. *Something New under the Sun: An Environmental History of the Twentieth-Century World.* New York: Norton, 2000.

Montgomery, John D., ed. *International Dimensions of Land Reform.* Boulder, Colo.: Westview Press, 1984.

Munasinghe, Mohan, and Jeffrey McNeely, eds. *Protected Area Economics and Policy: Linking Conservation and Sustainable Development.* Washington, D.C.: World Bank and World Conservation Union (IUCN), 1994.

Myers, Norman. "Threatened Biotas: 'Hot Spots' in Tropical Forests." *The Environmentalist* 8, no. 3 (1988): 187–208.

Nash, Roderick. *Wilderness and the American Mind*. New Haven, Conn.: Yale University Press, 1982.

Ness, Sally. *Where Asia Smiles: An Ethnography of Philippine Tourism*. Philadelphia: University of Pennsylvania Press, 2003.

Oposa, Antonio A., Jr. "A Socio-Cultural Approach to Environmental Law Compliance: A Philippine Scenario." Unpublished manuscript, 1997.

Owen, Norman G. *Prosperity without Progress: Manila Hemp and Material Life in the Colonial Philippines*. Quezon City: Ateneo de Manila Press, 1984.

Palawan Council for Sustainable Development, Office of the President. *Strategic Environmental Plan for Palawan: Towards Sustainable Development*. Manila: Palawan Integrated Area Development Project Office, [1989?].

Paredes, Ruby. "The Pardo de Taveras of Manila." In *An Anarchy of Families: State and Family in the Philippines*, edited by Alfred W. McCoy. Madison: Center for Southeast Asian Studies, University of Wisconsin-Madison, and Manila: Ateneo de Manila University Press, 1994.

Pauly, Daniel. "The Overfishing of Marine Resources: Socioeconomic Background in Southeast Asia." *Ambio* 17, no. 3 (1988): 200–206.

Pauly, Daniel, Villy Christensen, Rainer Froese, and Maria Lourdes Palomares. "Fishing Down Aquatic Food Webs." *American Scientist* 88 (January–February 2000): 46–51.

Pauly, Daniel, Villy Christensen, Sylvie Guénette, Tony J. Pitcher, U. Rashid Sumaila, Carl J. Walters, R. Watson, and Dirk Zeller. "Towards Sustainability in World Fisheries." *Nature* 418 (August 8, 2002): 689–695.

Philippine Campaign for Agrarian Reform and Rural Development (PhilCARRD). Proceedings of the Agrarian Reform and Rural Development National Consultation. Quezon City: Parnership for Agrarian Reform and Rural Development Services, Inc., June 1995.

Pollock Sturdevant, Saundra, and Brenda Stoltzfus. *Let the Good Times Roll: Prostitution and the U.S. Military in Asia*. New York: New Press, 1993.

Pomeroy, Robert S., and Melvin B. Carlos. "Community-Based Coastal Management in the Philippines: A Review and Evaluation of Programs and Projects, 1984–1994." *Marine Policy* 21, no. 5 (1997): 445–464.

Primack, Richard. *Primer of Conservation Biology*. Sunderland, Mass.: Sinauer Associates, 1995.

Putnam, Robert D. *Making Democracy Work: Civic Traditions in Modern Italy*. Princeton, N.J.: Princeton University Press, 1993.

Putzel, James. *A Captive Land: The Politics of Agrarian Reform in the Philippines*. Quezon City: Ateneo de Manila University Press, 1992.

Rasalan, Santos B. "Marine Fisheries of the Central Visayas." *The Philippine Journal of Fisheries*. Manila: Department of Agriculture and Natural Resources, 1957. Pp. 53–88.

Repetto, Robert. "Accounting for Environmental Assets." *Scientific American* (June 1992): 94–100.

Republic of the Philippines. Health Indicators. Department of Health. www.doh.gov.ph/healthsit/health_indicators.htm.

Republic of the Philippines. *Handbook for CARP Implementors.* 2nd ed. Quezon City: Department of Agrarian Reform, 1995.

Republic of the Philippines. *Medium Term Development Plan: 1993–1998.* Manila: National Economic and Development Authority, March 1995.

Republic of the Philippines. "Metro Manila Urban Transportation Integration Study." Draft final report. December 1998.

Republic of the Philippines. *1994 Philippine Yearbook.* Manila: National Statistics Office, 1995.

Republic of the Philippines. *The Philippine Economic Atlas.* Manila: Philippine Office of the President, [1964?].

Republic of the Philippines. *Philippine Fisheries Profile.* Quezon City: Bureau of Fisheries and Aquatic Resources, Department of Agriculture, 1997, 1998.

Republic of the Philippines. "Survey on Overseas Filipinos." Manila: National Statistics Office. www.census.gov.ph.

Richter, Linda K. *The Politics of Tourism in Asia.* Honolulu: University of Hawaii Press, 1989.

Rimban, Luz. "Big Companies Steal Water." Manila: Philippine Center for Investigative Journalism, March 6–7, 2000. www.pcij.org/stories/2000/water.html.

Rivera, Temario C. *Landlords and Capitalists: Class, Family, and State in Philippine Manufacturing.* Quezon City: University of the Philippines Press, 1994.

Rizal, José. *The Lost Eden (Noli Me Tangere).* Translated by León Ma. Guerrero. New York: Norton, 1961.

Roberts, Callum. "Effects of Fishing on the Ecosystem Structure of Coral Reefs." *Conservation Biology* 9 (October 1995): 988–995.

Rocamora, Joel. "Classes, Bosses, Goons, and Clans: Re-imagining Philippine Political Culture." In *Boss: Five Case Studies of Local Politics in the Philippines,* edited by Jose F. Lacaba. Manila: Philippine Center for Investigative Journalism and Institute for Popular Democracy, 1995.

Roque, Celso R. *Earth, Water, Air, Fire: Essays on Environmentalism.* Manila: Kalikasan Press, 1990.

Rush, James. *The Last Tree: Reclaiming the Environment in Tropical Asia.* New York: Asia Society and Westview Press, 1991.

Russ, Garry R., and Angel C. Alcala. "Do Marine Reserves Export Adult Fish Biomass? Evidence from Apo Island, Central Philippines." *Marine Ecology Progress Series* 132 (1996): 1–9.

Russ, Garry R., and Angel C. Alcala. "Effects of Intense Fishing Pressure on an Assemblage of Coral Reef Fishes." *Marine Ecology Progress Series* 56 (1989): 13–27.

Russ, Garry R., and Angel C. Alcala. "Marine Reserves: Rates and Patterns of Recovery and Decline of Large Predatory Fish." *Ecological Applications* 6, no. 3 (1996): 947–961.

Sale, Peter F., Graham E. Forrester, and Phillip S. Levin. "Reef Fish Management." *National Geographic Research and Exploration* 10, no. 2 (1994): 224–235.

Sandalo, Ricardo M. "Sustainable Development and the Environmental Plan for Palawan." In *Palawan at the Crossroads: Development and the Environment on a Philippine*

Frontier, edited by James F. Eder and Janet O. Fernandez. Manila: Ateneo de Manila University Press, 1996.

Schirmer, Daniel B., and Stephen Rosskamm Shalom, eds. *The Philippines Reader: A History of Colonialism, Neocolonialism, Dictatorship, and Resistance*. Boston: South End Press, 1987.

Scott, William Henry. *Barangay: Sixteenth-Century Philippine Culture and Society*. Quezon City: Ateneo de Manila University Press, 1994.

Serageldin, Ismail, Michael A. Cohen, and K. D. Sivaramakrishnan. "The Human Face of the Urban Environment." *Proceedings of the Second Annual World Bank Conference on Environmentally Sustainable Development*. Washington, D.C.: World Bank, 1994.

Steen, Harold K. *The U.S. Forest Service: A History*. Seattle: University of Washington Press, 1976.

Steinberg, David Joel. *The Philippines: A Singular and a Plural Place*. 4th ed. Boulder, Colo.: Westview Press, 2000.

Subida, Ronald, and Dan MacNamara. "Metropolitan Manila: A Focus on the Transport Sector," Integrated Environmental Strategies Philippines Project Report, Manila Observatory and USAID, June 2005, available at www.epa.gov/ies/documents/philippines/ies_update.pdf.

Sullivan, Rodney J. *Exemplar of Americanism: The Philippine Career of Dean C. Worcester*. Ann Arbor: University of Michigan Center for South and Southeast Asian Studies, 1991.

Tadem, Eduardo, Johnny Reyes, and Linda Susan Magno. *Showcases of Underdevelopment in Mindanao: Fishes, Forest, and Fruits*. Davao City: Alternate Resource Center, 1984.

Taylor, Harden F. "Survey of Marine Fisheries of North Carolina." Chapel Hill: University of North Carolina, 1951.

Terami-Wada, Motoe. "Karayuki-san of Manila: 1890–1920," *Philippine Studies* 34 (1986): 287–316.

Terborgh, John. *Diversity and the Tropical Rain Forest*. New York: Scientific American Library, 1992.

Totman, Conrad. *The Green Archipelago: Forestry in Pre-industrial Japan*. Berkeley: University of California Press, 1989.

Truong, Thanh-Dam. *Sex, Money, and Morality: Prostitution and Tourism in Southeast Asia*. London: Zed Books, 1990.

United Nations Department of International Economic and Social Affairs. *Population Growth and Policies in Mega-Cities: Metro Manila*. Population Policy Paper No. 5. New York: United Nations, 1986.

United States Bureau of Insular Affairs, War Department. *Reports of the Philippine Commission: 1900–1903*. Washington, D.C.: Government Printing Office, 1904.

Vitug, Marites Dañguilan. *Power from the Forest: The Politics of Logging*. Manila: Philippine Center for Investigative Journalism, 1993.

Wallace, Alfred Russel. "Equatorial Vegetation." *Tropical Nature and Other Essays*. 1878. Available at www.wku.edu/~smithch/wallace/S289.htm.

Wallace, Alfred Russel. "Physical Geography." Chap. 7 in *The Malay Archipelago*. Vol. 1. Available at www.worldwideschool.org.

Western, David, and R. Michael Wright. *Natural Connections: Perspectives in Community-Based Conservation.* Washington, D.C.: Island Press, 1994.

White, Alan T., and Annabelle Cruz-Trinidad. "The Values of Philippine Coastal Resources: Why Protection and Management Are Critical." Cebu City: Coastal Resource Management Project, 1998.

Wilson, Edward O. *Biophilia.* Cambridge, Mass.: Harvard University Press, 1984.

Worcester, Dean C. *The Philippines Past and Present.* New ed. New York: Macmillan, 1930.

World Bank. *Managing Fishery Resources.* Discussion Paper 217. Edited by Eduardo A. Loayza. Washington, D.C.: World Bank, 1994.

World Bank. *Philippines Environment Monitor,* 2002 and 2004. Washington, D.C., and Manila: World Bank, 2002 and 2004.

World Bank. *Urban Air Quality Management Strategy in Asia: Metro Manila Report.* World Bank Technical Paper No. 380. Washington, D.C.: World Bank, 1997.

World Bank Environment Department. *Republic of the Philippines: Conservation of Priority Protected Areas.* Project document. Washington, D.C.: World Bank, April 1994.

Worster, Donald, ed. *American Environmentalism: The Formative Period, 1860–1915.* New York: John Wiley and Sons, 1973.

Worster, Donald. *Nature's Economy: A History of Ecological Ideas.* 2nd ed. Cambridge: Cambridge University Press, 1994.

Zwick, Jim, ed. *Mark Twain's Weapons of Satire: Anti-Imperialist Writings on the Philippine-American War.* Syracuse, N.Y.: Syracuse University Press, 1992.

Index

Bolinao, Pangasinan, 111–113, 115–123,
124–130
access to work, 130
Bukidnon, Mindanao, 31, 62
Baungon, 74–79
Impasugong, 66–70
San Fernando, 30–40. *See also* Reforestation, San Fernando
Valencia, 42–43
Bulatao, Victor Gerardo, 159
Bureau of Fisheries, 89
Burnham, Daniel, 185

Cabrido, Candido, 124–125
CAFGU, 42, 43, 46, 60, 142–143
Canada, forest conservation, 50
Catholic Bishops Conference of the Philippines, 43
Catholic Church, 166. *See also* Gervais,
Charles; Kelly, Patrick; Satur, Nerilito
land use, 169
opposition to logging, 5, 7, 42–44
Cebu, 99, 111
Cement industry, 111, 116
controversy in Bolinao, 111–113, 115–123,
124–130
Central Luzon, 113–114
Central Visayas, 99
Christian and Muslim conflicts, 59
Civil society. *See* NGOs
Clean Air Act, 182
Clear-cutting, 52
costs of, 28–29, 52–53. *See also* Tropical
forests; Logging
Cleveland, Ohio, 122, 186
Coastal decline, costs of, 127–128. *See also*
Fisheries, decline; Overfishing
Coastal management, 86–87, 89–90, 93–97,
99–103, 106–108
community role, 91–92, 99
livelihood, 101–103, 106–109
local government, 100
scientific basis, 96–97
Coastal Resource Management Project
(CRMP), 99–103, 107–109
Coastal villages, standard of living, 89
Coastline, Philippine, 85
Coasts, 85–86
development, 86, 122
economic value, 85–86. *See also* Fisheries, exports
pollution, 86

population, 85–86
Cogon grass, 30, 62, 88
Cojuangco family, 170
Colonialism and dependence, 123
Commercial logging. *See* Logging industry
Community organizing, 76–79, 92–93, 116
Comprehensive Agrarian Reform Program
(CARP), 157–160, 162–164, 171–177
agrarian reform communities, 160
effects on investment, 171
Conservation. *See also* Environmentalism;
Natural resource management
attitudes toward, 88
of Canadian forests, 50
community role, 73–74, 75, 99, 112, 116,
143
debt-for-nature swap, 139–140, 144
development and, 153, 209–212
economic growth and, 209. *See also*
Development; Economics
enforcement, 140–145
Europe, 52
foreign funding, 101
forests, 48–53
government and, 50, 53, 62, 86, 87,
89–90, 91, 93, 95, 99,100–109, 143
history, 48–53, 70–74
industrialization and, 108–109
international, 70–73, 77
Japan, 58
livelihood and, 63–64, 67–70, 77–78,
80–82, 101, 145,
local conflicts, 137
obstacles to, 63–64, 74–79, 80–82, 99
Philippine forests, 52. *See also* Forests,
Philippines
versus preservation, 50
private property and, 50, 99
protected areas. *See* Protected areas
restricted access, 99
stakeholders, 145–146
successes in, 86
United States, 48–51, 52, 58
violence and, 41–46, 144–145. *See also*
Logging, military and
Conservation of Priority Protected Areas
Project (CPPAP), 62–65, 74–82
livelihood and, 67–70
Constitution, 189, 192
Convention on Biological Diversity, 72
Copper mining, 118. *See also* Marinduque
Coral reef fisheries. *See* Reefs; Reef fish

Trade, 164
  of birds' nests, 165
  logging and, 60
  regional, 48, 115, 116, 117
  United States and Philippine, 165–166.
    See also United States, relations with
    Philippines
Traditional politicians, 104–106
Transportation, 181
Trapos. See Traditional politicians
Trawlers, 89
Tropical forests, 26–29. *See also* Forests
  fast-growing species, 38, 68
  loss, 72
  soil, 28–29. *See also* Clear-cutting;
    Deforestation; Logging
Tsismis, 21–22
Typhoon Uring, 19–21. *See also* Ormoc,
  Leyte

Underdevelopment, 111, 115, 176
United States
  Anti-Imperialist League, 47
  as colonial power, 47, 168–169
  conservation, 48–51, 52, 58
  environmental laws, 56
  Forest Service, 51
  HIV/AIDS and, 147
  insular government, 47, 51, 185. *See also*
    Philippine Commission
  logging industry, 48, 49, 51–52
  military and prostitution, 147
  military bases, 124
  national forests, 50, 58
  natural resource use, 48–51, 57–58
  relations with Philippines, 47–48, 50–52,
    91, 165–166, 170
  relations with Marcos, 170
  urban renewal, 186
  wood industry. *See* United States, log-
    ging industry
U.S. Agency for International Development
  (USAID), 91
University of the Philippines, 118
  Los Baños, 51

Marine Science Institute, 86, 91, 116–118,
  126
Urban development. *See* Development,
  urban
Urban environment, 178, 181, 201
  income disparities and, 180, 202
Utang na loób, 21, 55

Valencia, Bukidnon, 42–43
Vietnamese refugees, 133, 151
Vitug, Marites Dañguilan, 56
von Humboldt, Alexander, 25

Wallace, Alfred Russel, 25
Warlords, 42, 137
Water, privatization. *See* Metro Manila,
  water privatization
  use, 117
Water companies, 203–206. *See also* Manila
  Water Co.; Maynilad Water
  "nonrevenue losses," 204
Water Crisis Act, 202
Watershed management, 24, 49, 58, 201
Water supply, global, 205
  urban, 178, 180, 201–206
Western, David, 73
White, Alan, 91, 93–94, 99
Wilderness, 50, 61, 70–71
Wilson, Edward O., 71
Wood industry. *See* Logging industry
Worcester, Dean C., 47–49, 52, 58, 119
World Bank, 62, 80, 202
World Conservation Union. *See* IUCN
World War II, aftermath, 52, 54, 186,
  187
  occupation, Japan. *See* Japan, World
    War II occupation
World Wildlife Fund, 139
Worster, Donald, 71

Yabes, Criselda, 153
Yellowstone, 70–71
Yosemite, 70

Zambales Range, 115